T0135555

Algebraic Multigrid for the Multi-Ion Transport and Reaction Model

-

A Physics-Aware Approach

Inaugural – Dissertation
zur Erlangung des Doktorgrades der
Mathematisch-Naturwissenschaftlichen Fakultät der
Universität zu Köln

vorgelegt von
Peter Thum
aus Köln

2012

Berichterstatter:
Prof. Dr. Ulrich Trottenberg
Prof. Dr. Caren Tischendorf

Tag der mündlichen Prüfung: 05.11.2012

Bibliografische Information der Deutschen Nationalbibliothek

Die Deutsche Nationalbibliothek verzeichnet diese Publikation in der
Deutschen Nationalbibliografie; detaillierte bibliografische Daten sind
im Internet über http://dnb.d-nb.de abrufbar.

ISBN 978-3-8325-3285-7

Logos Verlag Berlin GmbH
Comeniushof, Gubener Str. 47,
10243 Berlin
Tel.: +49 (0)30 42 85 10 90
Fax: +49 (0)30 42 85 10 92
INTERNET: http://www.logos-verlag.de

Zusammenfassung

In vielen Herstellungsprozessen werden heutzutage elektrochemische Verfahren eingesetzt. Diese Prozesse werden meist vorab durch eine Simulation optimiert, um den Verbrauch von Ressourcen zu minimieren. Hierzu wird das Multi-Ionen Transport und Reaktionsmodell (MITReM) eingesetzt. Das Modell ist ein nichtlineares System von partiellen Differenzialgleichungen (PDEs). Es wird durch eine kombinierte Residuen-Distribution Finite Elemente Methode diskretisiert und mit einem Newton Verfahren linearisiert, dies resultiert in einer Reihe von großen linearen Gleichungssystemen.

In dieser Arbeit wollen wir diese linearen Systeme durch ein punktbasiertes algebraisches Mehrgitterverfahren (PAMG) lösen. Diese Art von Verfahren ist bekannt für ihre optimale Skalierbarkeit für eine Reihe von Problemklassen. Wir zeigen, dass standardmäßige PAMG Verfahren, die auftretenden linearen Systeme nicht zufriedenstellend lösen, weil physikalische Eigenschaften des MITReMs nur unzureichend berücksichtigt werden.

Wir entwickeln ein neues PAMG Verfahren, das physikalische Eigenschaften eines zugrunde liegenden PDE Systems explizit mit einbezieht. Das physikalisch-orientierte Verfahren löst die entstehenden Gleichungssysteme effizient und robust.

Das neue PAMG Verfahren berücksichtigt die physikalischen Eigenschaften sowohl in der Glättung als auch in der Grobgitter Korrektur. Das Glättungsverfahren verwendet eine auf physikalischen Eigenschaften basierende Permutation der Variablen, um bessere Ergebnisse zu erzielen. Die Permutation wird durch eine automatische Methodik bestimmt, die sowohl skalare PDEs als auch PDE Systeme umfasst und insbesondere Informationen über Richtungsfelder mit einbezieht.

Wir entwickeln eine heuristische Péclet Zahl, die es möglich macht numerische Schwierigkeiten auf groben Gittern zu lokalisieren. Diese Schwierigkeiten werden durch die verwendete Diskretisierung verursacht. Das vorgeschlagene Grobgitter Korrekturverfahren ermöglicht, diese Erkenntnisse in den Vergröberungsprozess mit einzubeziehen.

Das neue physikalisch-orientierte PAMG Verfahren vergleichen wir mit standardmäßigen PAMG Verfahren an der Konvektions-Diffusionsgleichung, an einem Migrations-Diffusionssystem und an mehreren industriell relevanten Geometrien und elektrochemischen Systemen des MITReMs. Die Vergleiche zeigen, dass das neue Verfahren den standardmäßigen PAMG Verfahren in Effizienz und Robustheit überlegen ist.

Abstract

Electrochemical processes are essential in the manufacturing of a very broad range of products. The multi-ion transport and reaction model (MITReM) is used to simulate such electrochemical processes. The governing partial differential equations (PDEs) are discretized in space by a combined finite element and residual-distribution scheme. The discrete system is linearized with Newton's method which results in a series of linear systems. The focus of this work is on the development of a stable and efficient point-based algebraic multigrid (PAMG) method for the solution of the linear systems which arise in the MITReM simulation.

The new PAMG approach makes use of physics-aware smoothing and coarse-grid correction techniques. This means that it, explicitly, takes important physical properties of the PDE system into account.

We introduce a reordering framework which makes it possible to use physics-oriented matrix-based measures for strength of connectivity to derive application-specific point orderings for smoothing. The framework's generality not only allows the determination of suited permutations of the variables for the electrochemical simulations considered, but also for other convection- and/or migration-dominated simulation tasks.

We introduce a heuristic Péclet number to locate areas causing numerical difficulties within the hierarchy of the algebraic multigrid (AMG) method for scalar PDEs as well as systems of PDEs. We investigate several coarse-grid correction techniques which take this information into account and show results for the convection-diffusion equation, the migration-diffusion system and the MITReM.

We motivate the use of the physics-aware PAMG components and apply the approach for a range of geometries and chemical setups with scientific and industrial relevance. Our numerical experiments illustrate, in particular, that physical properties of the underlying problem have to explicitly be considered for constructing efficient and robust AMG-based methods.

Contents

Chapter 1

Introduction

> The only interesting fields of science are the ones where you
> still don't know what you're talking about.

> - I.I. Rabi

1.1   Context

In today's industry, numerical simulation is used to accelerate principal
design decisions and prototyping. Based on the advances in modeling,
computer technology and in numerical methods, numerical simulation is
nowadays able to solve many complex engineering problems and realisti-
cally model industrial processes and physical processes occurring in nature.
Furthermore, numerical simulation also helps to get a feeling for natural
phenomena which cannot be tested by experiments.

The underlying physical behavior and properties of such processes have
to be modeled in order to simulate those overall processes and systems. For
a large class of problems, this is done with partial differential equations
(PDEs) derived from balance equations. The arising models, which usually
consist of various coupled PDEs which build up a system, are discretized.
Hereby, they are usually transformed into equations for a finite number of
points, which are then solved approximately.

The simulations shall be as realistic as possible to model the real world
physical effects. Hence, not only the number of points used to discretize
the systems of PDEs is large, also macro- and even microscopic effects are
considered. For many applications, this results in discretized systems with

1

several millions of variables. In hand with parallelization and increasing computer memory the accuracy of the virtual representation of reality and also the size of the discretized systems increase. The discretized systems are usually non-linear and/or time-dependent. As a consequence, the systems are linearized and/or a series of states has to be solved. That is, in order to find an approximate solution for a model, a large number of huge linear systems has to be solved.

In this work, we consider the simulation of electrochemical processes. These processes are, e.g., used for decoration, functional coating (wear, corrosion, morphology), electro forming (EF), and electrochemical machining (ECM). High-end applications are found in electronics (e.g., plating on wafers and PCBs), aeronautics (coatings, ECM and EF of complicated parts such as turbine blades with micro-cooling channels and molds), medicine (needles, implants), automotive (motors, injection systems), gas production, etc. Another very important family of electrochemical processes is found in corrosion. In these applications, the process is driven by differences in local properties (material, environment).

The electrochemical processes considered are modeled using the multi-ion transport and reaction model (MITReM). The MITReM incorporates the transport equation of each ion, the electro-neutrality constraint, and the non-linear Butler-Volmer boundary conditions describing the electrochemical processes at the electrodes. Typically, the total number of equations of the PDE system is between 3 and 10.

The MITReM is discretized with a combined residual-distribution finite element method, resulting in large sparse matrix equations. Sparse means that the resulting matrices have only few (typically 15-45 in the case of the MITReM) non-zero entries per matrix row. The arising discretized system is linearized with a Newton method. Accordingly, several linear systems have to be solved during a simulation. This results in a very high amount of work spent to solve the linear systems compared to the overall simulation time (usually more than 80% using standard solvers). For this reason, the arising linear systems have to be solved as fast and efficiently as possible in order to perform a simulation run in an appropriate time.

One way to solve these linear systems is to use state-of-the-art direct solvers like PARDISO [53] or MUMPS [1]. However, such kind of solvers need a high amount of memory and thus are not applicable for very large systems of equations. Therefore, iterative solvers, which commonly need less memory, are often used to solve such systems. Frequently, iterative solvers employed are ILU-preconditioned Krylov methods since they are very robust. Such a method is also used in the simulation code considered. However, these methods have the disadvantage that their convergence rate usually decreases with increasing number of variables. Hence, the time spent solving such equations grows super-linearly with the size of the matrices.

Another very important class of methods to solve sparse linear systems is based on multigrid/level techniques. Such methods are characterized by their linear complexity for a large class of problems. To be more specific, the run time to solve a linear system rises only proportionally with the matrix size. This property is called scalability or optimal complexity. Hence, we investigate if such methods can also be used to solve the linear systems which arise for the MITReM.

We consider algebraic multigrid (AMG). AMG provides a state-of-the-art technology to develop solvers for large sparse linear systems with high efficiency. AMG's major advantages - not only numerical efficiency, but also robustness, scalability and ease-of-use - have become the driving forces behind its growing success as numerical kernel in many industrial simulation codes. AMG is used in diverse areas such as fluid flow, structural mechanics, oil reservoir and groundwater simulation, casting and molding, process and device simulation in solid state physics, and circuit simulation.

However, considering convection-dominated problems, standard AMG approaches do not usually exhibit optimal complexity. For complicated applications, like the electrochemical systems considered, which are of convection-migration-diffusion-reaction type even divergence occurs for industrially relevant model settings.

1.2 Main Focus and Structure

We consider the point-based algebraic multigrid (PAMG) strategy which has been developed in [90] for strongly coupled systems of PDEs such as the MITReM. This strategy has been extended and successfully applied to drift-diffusion systems from device simulation [29]. Since the MITReM is composed of convection, diffusion, reaction and migration (which is essentially the same as drift), we have to consider two new aspects in the context of the MITReM. First, we have to deal with convection and reaction terms (in a form different from device simulation). Second, in most industrial applications, migration and convection directions are nearly orthogonal to each other, which makes it difficult to create an optimal PAMG approach.

We compare the one-level iterative solver employed in the simulation software to PAMG and show that such multigrid solvers can drastically (up to 10 times) reduce the run time of the simulation software. Furthermore, we demonstrate that state-of-the-art PAMG approaches are either very robust, but not fast, or fast, but not robust, where "not robust" means that they diverge for industrially relevant steady-state MITReM simulations. There are many reasons for this which will be discussed in detail in this work. However, all difficulties have in common that they occur because physical properties of the PDE system are not represented well in the standard PAMG framework.

Hence, we develop a physics-aware multigrid approach and show that it is robust and efficient for industrially relevant steady-state MITReM simulations. Physics-aware means that the approach, explicitly, makes use of physical information of the PDE system, like the convection-direction or the direction of the electromagnetic field. The new approach will be integrated in the linear solver library SAMG [37] from the Fraunhofer Institute for Algorithms and Scientific Computing (SCAI) which includes state-of-the-art AMG technology.

We briefly describe the physics-aware multigrid approach, here. Just like all multigrid/level methods, the PAMG strategy makes use of two principles which have to be well attuned to one another - **smoothing** and **coarse-grid correction**. Hence, there are two possibilities in order to create a stable and efficient PAMG method. One possibility is to adjust the smoother. The other possibility is to improve the coarse-grid correction of the multigrid method. However, improving just one of the components does not lead to a robust and efficient approach for the MITReM. Hence, both components have to explicitly take physical properties of the PDE system into account.

The MITReM belongs to the class of convection-dominated problems. For these type of problems, it is well known that the ordering of grid points used within the solution process drastically influences the convergence rate of iterative one-level methods. Furthermore, the smoother of the AMG method has not only to smoothen the error, but also to reduce the "convective" part of the error. Since the smoothers employed are iterative one-level methods, the ordering of grid points has a strong influence on AMG's convergence rate [62, and references therein].

Suitable orderings are usually computed with the help of the original matrix. Sometimes, information from the discretization is used. However, it has not been investigated so far in which way explicitly given physical information can be employed to create such orderings. Furthermore, the possibility to reorder **systems of PDEs** in the context of AMG has not been investigated so far. Additionally, it is not clear how the two **different directions of movement** (migration and convection) in the case of the MITReM can be handled appropriately.

We develop a **reordering framework** which makes it possible to employ explicitly given physical information and physics-oriented matrix-based measures for strength of connectivity to derive **application-specific point orderings** for smoothing. The reordering framework provides various settings for creating permutations which can be applied independently for each level. Therefore, it is flexible enough to meet the demands of a physics-oriented smoothing within an AMG method not only for the electrochemical simulations considered, but also for other convection- and/or migration-dominated simulation tasks.

Besides the physics-aware smoothing procedure, we create a **physics-aware coarse-grid correction**. The physics-aware coarse-grid correction enables us to take certain physical properties of the PDE system into account which are not present or not represented well in the linearized systems. Considering the MITReM, two kinds of problems arise within the coarse-grid correction which we resolve with the physics-aware approach. These problems are caused by both the **nonlinearity** and the choice of the **discretization**.

The nonlinearity of the system, in combination with the first guess of the Newton method, leads to an underrepresentation of physical effects described by migration in the beginning of the Newton method. This causes a bad convergence behavior or even divergence for linear systems from early Newton steps using state-of-the-art PAMG approaches. The physics-aware coarse-grid correction developed shows much better results because it is able to **stress the relevance of the migration** in the coarse-grid correction procedure by using external information like the position of the electrodes.

Considering finite element discretization of convective terms, it is well known that linear shape functions will lead to checkerboard instabilities if the **Péclet condition** is not fulfilled [50]. In order to fulfill the Péclet condition, the mesh has to be fine enough. Another possibility to avoid checkerboard instabilities is to stabilize or change the discretization, e.g., by using the residual-distribution scheme.

The MITReM is discretized via a combined residual-distribution finite element method. That is, the convection is discretized via the residual-distribution scheme [83]. The other parts of the MITReM are discretized via finite elements using linear shape functions [12]. However, the migration term numerically behaves very similarly to the convection term. Hence, using an insufficiently refined grid, checkerboard instabilities may arise for the discretized migration term.

Furthermore, in the context of geometric multigrid, it is well-known that, even if the Péclet condition is fulfilled on the fine grid, it does not have to be fulfilled on all coarser grids [106]. We show that this is not the case if classical AMG is used.

We introduce a **heuristic Péclet number** for the MITReM which takes the convection, migration and diffusion of the PDE system into account. The heuristic Péclet number is **purely based on algebraic information**, and does not explicitly use grid information.

We show that the heuristic Péclet condition is locally violated for the electrochemical setups considered. Unfortunately, this cannot be remedied by using a finer grid resolution because of memory restrictions of the computing architecture. However, the quality of the solution of the Newton method nonetheless fulfills the accuracy-requirements of the engineers, hence, the discretization type is not changed.

We investigate the effects of a locally violated Péclet condition for the hierarchy of AMG. Since the heuristic Péclet number introduced is purely based on algebraic criteria, we can also compute it on coarse levels where no grid information is available anymore. We observe that the Péclet numbers increase and that the area of violated Péclet condition spreads out on coarse levels when using standard AMG components. The heuristic Péclet number is used to **localize the numerical difficulties** on all levels of the AMG method. The gained information is then employed to construct a physics-aware coarsening and interpolation.

The new AMG components form a physics-aware AMG approach which can be used to solve various types of problems and, especially, the linear systems from the MITReM. We apply the new physics-aware AMG approach to a range of model problems of convection-diffusion and migration-diffusion type to show its benefits compared to the standard AMG. Finally, we apply the physics-aware PAMG approach to solve the MITReM and perform numerical experiments for a range of industrially relevant electrochemical problems to demonstrate the robustness and efficiency of the PAMG approach newly designed.

The remainder of this work is structured as follows.

Chapter 2 introduces the MITReM. It provides a description of the electrochemical physics involved, followed by a description of the discretization of the model.

Chapter 3 provides a brief introduction to AMG. It focuses on classical AMG and its extension PAMG to strongly coupled systems of PDEs. Furthermore, it summarizes state-of-the-art techniques used in multigrid methods in order to solve convection-dominated flow problems.

Chapter 4 introduces the reordering framework developed and used to build a physics-oriented smoother. It outlines how underlying physical properties, e.g., direction-dependency of convection, of a linear system are taken into account by the reordering framework.

Chapter 5 derives the heuristic Péclet number for scalar equations. Based on this, the heuristic Péclet number for systems of PDEs is introduced which takes convection, diffusion and migration into account. The heuristic Péclet number makes it possible to localize numerical problems on each level of the AMG hierarchy and to develop a coarse-grid correction which takes this information into account. Several physics-aware coarsening techniques are investigated for a convection-diffusion model problem.

Chapter 6 presents numerical results for the application of physics-aware coarsening in the case of a migration-diffusion system. The model problem considered is derived from the MITReM and gives insight which physical coarsening techniques are suited best for the MITReM.

Chapter 7 introduces a physics-aware PAMG approach for the MITReM. This approach makes use of the new techniques. It presents numerical results for the new approach and compares it to state-of-the-art PAMG. Furthermore, a comparison between the standard one-level iterative solver of the simulation software and PAMG is provided. The numerical experiments illustrate the robustness and efficiency of the new methods.

Finally, **Chapter** 8 concludes this work and gives an outlook on further research.

Remark 1.1. *Parts of this work have been published in [104, 105].*

Notation 1.2. *Throughout this thesis bold letters always denote vectors or matrices.*

Chapter 2

The Multi-Ion Transport and Reaction Model (MITReM)

We introduce the multi-ion transport and reaction model (MITReM) and its discretization. The model describes the effects of convection, diffusion, migration, and chemical as well as electrode reactions on the concentration, and potential distributions in an electrochemical system.

An electrochemical system consists of electrodes which are in contact with an electrolyte and connected through an external electronic conductor (outer circuit). An electrolyte is a substance in which the mobile species are ions and the free movement of electrons is blocked. Figure 2.1 shows a sketch of an electrochemical system. Note that the inflow and outflow are usually positioned opposite to each other.

Remark 2.1. *We use the expression species to refer to ions as well as neutral molecular components that do not dissociate.*

The electrolyte is circulating in order to ensure a uniform distribution of the solved species. Additionally, the flow of electrons in the outer circuit causes a flow of ions in the electrolyte. This means, there is a transfer of material and charge in the electrolyte. Hence, a mathematical model of the mass transfer of ions from the solution to the surface of the electrodes requires a description of the mobile ionic species based on material balances, energy balances, electro-neutrality and fluid flow. In the most general case, it is necessary to combine all these phenomena in order to calculate the mass and charge transfer in electrochemical reactors. Since

9

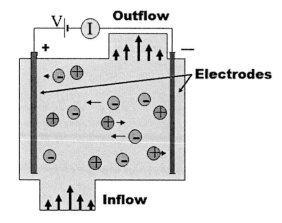

Figure 2.1: Exemplary layout of an electrochemical system.

this is in practice nearly impossible, simplifications to the model are used today. For instance, when dealing with dilute solutions, the cross diffusion and a different velocity for each species is not considered in the transport equations. Furthermore, it is assumed that the concentration of the species does not affect the velocity field that is determined by the solvent (usually water).

At the electrodes, electrochemical reactions take place. In contrast to a chemical reaction, electrochemical reactions involve at least one outer sphere electron transfer, by which an overall chemical change occurs. The overall reaction taking place in an electrochemical system is composed of two independent half-reactions. This means, that reduction and oxidation occur at different places while in a chemical reaction, both, reduction and oxidation occur at the same place. The rate of the reaction can be controlled by the applied potential difference between the electrodes through the outer circuit. Finally, electrochemical reactions are **heterogeneous**, which means that they always occur at the interface between the electrolyte and the electrode. They are described by non-linear boundary conditions. The chemical redox reactions, which are also modeled by the MITReM, are called **homogeneous** reactions.

Details on the electrochemistry can be found in the textbooks [7, 77]. The MITReM has been implemented and validated in [8, 9, 108, 109] for laminar flow and one-dimensional, planar and axis-symmetrical configurations, and a wide variety of electrochemical systems. Furthermore, the

model has been validated and used for turbulent flows and mass transfer in
[75, 76].

In Section 2.1, we give a brief introduction to the physical background
of electrochemical systems, more information can be found in [77, and references therein]. In Section 2.2, we describe the homogeneous reactions. We
summarize the MITReM partial differential equation (PDE) system in Section 2.3. In order to finalize the basic description of the MITReM system,
we give a survey on the boundary conditions employed which also describe
the heterogeneous reactions in Section 2.4. In Section 2.5, we summarize
known theoretical results regarding the MITReM (existence and uniqueness
of solutions). Finally, we briefly describe the discretization of the electrochemical system employed and present links to the literature with details
in Section 2.6.

2.1   Electrochemical Background

The heterogeneous reactions are broken into two half-reactions at different
electrodes. Nevertheless, the *rates* of reaction are coupled. For a medium
of uniform dielectricity, this coupling is described by the conservation of
charge, $\mathrm{div}\mathbf{j} = 0$, where $\mathbf{j}$ is the current density, and Poisson's equation

$$\nabla^2 \Phi = -\frac{F}{\epsilon} \sum_i z_i c_i, \qquad (2.1)$$

with Coulomb potential Φ, Faraday constant F (96,485.309C/mol), and
uniform permitivity ϵ (ϵ depends on the electrolyte and its temperature -
size of 10^{-15}). $\sum_i z_i c_i$ is the sum over all species i in the solution with
c_i and z_i being the concentration and the charge number of the species
i, respectively. Note that the constant $\frac{F}{\epsilon}$ is so large that an appreciable
separation of charge would require prohibitively large electric forces. This
is not given for an electrolyte because the conductivity is so large that free
charges do not exist. Therefore, Poisson's equation is often replaced by the
electroneutrality equation [77],

$$\sum_i z_i c_i = 0. \qquad (2.2)$$

The imposition of a potential difference across an electronic conductor creates a driving force for the flow of electrons (electromagnetic field).
The electromagnetic field $\mathbf{E}$ is described by the vector of the electric field
strength (in V/m). Since the electromagnetic field is free of rotation it can

be written as a gradient of a scalar field. Namely, it can be written as the gradient in the Coulomb potential

$$\mathbf{E} = -\nabla\Phi. \tag{2.3}$$

Ohm's law describes the linear relation between the voltage U and the current strength I, which is valid for many electric conductors at constant and not too high/low temperature. The constant which describes their relation is called resistivity (independent of U and I). In the differential form it writes

$$\mathbf{j} = \sigma\mathbf{E}, \tag{2.4}$$

with $\mathbf{j}$ being the current density (in A/m^2) and $\sigma = 1/\rho$ the electric conductivity (with ρ specific resistivity) [93].

The current in a solution is the net flux of charged species (ions)

$$\mathbf{j} = F\sum_i z_i \mathbf{N_i}, \tag{2.5}$$

where F is the Faraday constant, $\mathbf{N_i}$ is the flux density (in mol/m^2s), and z_i is the charge of species i.

In contrast to a conductor, ions in an electrolyte move not only in response to an electromagnetic field (migration/drift), but also in response to concentration gradients (diffusion), and bulk fluid motion (convection). Therefore, the net flux of an ion is the sum of the migration, diffusion and convection terms. In the following, we take a closer look at each of these terms.

We consider electrodes in a solution with an electromagnetic field applied between them. The electromagnetic field creates a driving force for the motion of the charged species. It drives cations toward the cathode and anions toward the anode. That is, cations move in the direction opposite to the gradient in potential. The velocity of the ion in response to an electromagnetic field is its *migration* velocity, given by

$$\mathbf{v}_{i,\mathrm{mig}} = -Fz_i u_i \nabla\Phi, \tag{2.6}$$

where u_i is the mechanical mobility (in $m^2 mol/Js$). The mechanical mobility is commonly defined to be the ratio between mean particle velocity and a driving force, in the case considered, the electromagnetic field. The flux density of a species is equal to its velocity multiplied by its concentration. Hence, the flux density share of the migration is given by

$$\mathbf{N}_{i,\mathrm{mig}} = -Fz_i u_i c_i \nabla\Phi. \tag{2.7}$$

Besides migration, any concentration gradients drive mass transport by the process of diffusion. The component of the flux density of a species describing the diffusion is

$$\mathbf{N}_{i,\text{diff}} = -D_i \nabla c_i, \tag{2.8}$$

where D_i is the diffusion coefficient of species i.

Convection is the bulk movement of the fluid. The flux density of a species describing the convection is given by

$$\mathbf{N}_{i,\text{conv}} = c_i \mathbf{v}, \tag{2.9}$$

where $\mathbf{v}$ is the velocity field of the bulk fluid.

Concluding, we are able to describe the net flux density of a species given by the combination of the migration, diffusion and convection terms,

$$\mathbf{N}_i = -F z_i u_i c_i \nabla \Phi - D_i \nabla c_i + c_i \mathbf{v}. \tag{2.10}$$

Substitution into Equation (2.5) for the current gives

$$
\begin{aligned}
\mathbf{j} &= F \sum_i z_i \mathbf{N}_i \\
&= -F \sum_i z_i (F z_i u_i c_i \nabla \Phi + D_i \nabla c_i - c_i \mathbf{v}) \\
&= -F^2 \sum_i z_i^2 u_i c_i \nabla \Phi - F \sum_i z_i D_i \nabla c_i + \underbrace{F \sum_i z_i c_i \mathbf{v}}_{=0 \text{ (electroneutrality eq. } \sum_i z_i c_i = 0)} \\
&= -F^2 \sum_i z_i^2 u_i c_i \nabla \Phi - F \sum_i z_i D_i \nabla c_i.
\end{aligned}
$$
$$\tag{2.11}$$

We see that the bulk convection will not contribute to the net current, if we use the electroneutrality equation (2.2). If we use Poisson's equation (2.1), instead, the influence of the convection will be negligible, too. However, convection can cause mixing of the solution. In this way, fluid motion can affect concentration profiles and serve as an effective means to bring reactants to the electrode surface.

When there are no concentration variations in the solution, the equation reduces to the differential form of Ohm's law

$$\mathbf{j} = -\sigma \nabla \Phi, \tag{2.12}$$

where $\sigma = F^2 \sum_i z_i^2 u_i c_i$ is the conductivity of the solution. This is an expression of Ohm's law, valid for electrolytes in absence of concentration gradients. When there are concentration gradients, the current density is no longer proportional to the electromagnetic field, and Ohm's law does not hold.

Considering the diffusion term in Equation (2.11), we find that the current density could even have a different direction from the electromagnetic field. Solving Equation (2.11) for $\nabla\Phi$, we get

$$\nabla\Phi = -\frac{\mathbf{j}}{\sigma} - \frac{F}{\sigma}\sum_i z_i D_i \nabla c_i. \tag{2.13}$$

Hence, it is obvious that even in absence of a total or net current, there may be a potential difference caused by diffusion.

2.2   Homogeneous Reactions

Homogeneous reactions are elementary reactions which occur between different species. For simplicity, we only describe monomolecular and bimolecular reactions, which commonly occur, further details on homogeneous reactions can be found in [2].

Mono- or unimolecular reactions are reactions with only one reacting species, for example,

$$A \xrightarrow{k_1} B + C. \tag{2.14}$$

Here, species A reacts to species B and C at the rate k_1. This monomolecular reaction implies the rate law

$$\frac{\partial(c_B + c_C)}{\partial t} = k_1 c_A, \tag{2.15}$$

which is equivalent to

$$\frac{\partial(c_A)}{\partial t} = -k_1 c_A. \tag{2.16}$$

Note that the rules of stoichiometry hold. That is, if one reactant is being used up, an equivalent amount of the product is gained. In the case that the described reaction is reversible, the reversed reaction would be bimolecular.

Bimolecular reactions are, as the name implies, reactions with two reacting species. The reactions which usually occur in the MITReM are bimolecular in one direction and monomolecular in the reverse direction. That is, we typically have

$$A \underset{k_1}{\overset{k_2}{\rightleftharpoons}} B + C, \tag{2.17}$$

which implies the rate law

$$\frac{\partial c_A}{\partial t} = k_2 c_B c_C - k_1 c_A, \tag{2.18}$$

or equivalently

$$\frac{\partial(c_B + c_C)}{\partial t} = -k_2 c_B c_C + k_1 c_A. \tag{2.19}$$

Eventually, the system might reach its equilibrium, which means that the concentration of the species does not change in time anymore because their production and reduction rates are equal. In equilibrium,

$$\frac{\partial c_A}{\partial t} = \frac{\partial(c_B + c_C)}{\partial t} = 0 \tag{2.20}$$

holds. This implies

$$K := \frac{k_2}{k_1} = \frac{c_A}{c_B c_C}, \tag{2.21}$$

where K is called equilibrium constant.

In the MITReM, the change of concentration of species i caused by homogeneous reactions is denoted by R_i. Considering Equation (2.17), R_A would be $R_A = k_2 c_B c_C - k_1 c_A$.

2.3   Basic System

In the following, we summarize the four equations describing the MITReM. For every species i, the concentration change in each point of the solution is equal to the net input plus the local production caused by homogeneous reactions

$$\frac{\partial c_i}{\partial t} = -\text{div}(\mathbf{N}_i) + R_i. \tag{2.22}$$

The production per unit volume, R_i, involves homogeneous chemical reactions in the bulk of the solution, but not heterogeneous reactions, which are described by the electrode boundary conditions. In electrochemical systems, homogeneous reactions do not frequently influence the solution strongly and are, hence, ignored in which case R_i is zero.

The second equation of the MITReM considered is Poisson's equation (see Equation (2.1)). It describes the charge density at each point of the solution and is composed of the algebraic sum of the charges of all dissolved particles

$$-\nabla^2 \Phi = \frac{F}{\epsilon} \sum_i z_i c_i. \tag{2.23}$$

When the charge is zero at each point, the solution is said to be electroneutral. For an electrolyte, the conductivity is so large that free charges do not exist. Hence, the Poisson equation can be replaced by the electroneutrality equation, see also Section 2.1.

Note, the conservation of charge $\text{div}\mathbf{j} = 0$ can be derived by multiplying Equation (2.22) with $z_i F$ and using the electroneutrality equation as well as the fact that for homogeneous reactions the term $\sum_i z_i R_i$ vanishes in the bulk solution.

The missing two equations describe the fluid flow. First of all, there is the continuity equation

$$\text{div}\, \mathbf{v} = 0, \tag{2.24}$$

which implies that the density of the fluid is constant and not effected by concentration differences of dissolved species. In practice, this is usually the case. However, dependent on the model and chemical system also density differences may occur. In the case of different densities,

$$\frac{\partial \rho}{\partial t} = -\text{div}(p\mathbf{v}) \tag{2.25}$$

is used, where ρ is the density of the fluid.

The velocity field $\mathbf{v}$ is specified by the Navier-Stokes equation. The Navier-Stokes equation describes the conservation of momentum for incompressible Newtonian fluids. Therefore, it is often referred as momentum equation. The momentum equation writes

$$\frac{\partial \mathbf{v}}{\partial t} + (\mathbf{v}\nabla)\mathbf{v} = \frac{\nabla p}{\rho} + \nu\nabla^2\mathbf{v} + \mathbf{g}, \tag{2.26}$$

where p denotes the pressure, ρ the density and $\nu = \mu/\rho$ the kinematic viscosity of the fluid (μ denotes the viscosity of the fluid), and $\mathbf{g}$ the acceleration caused by gravitation.

Summarizing, the MITReM considered PDE system can be written as

$$\frac{\partial \mathbf{v}}{\partial t} + (\mathbf{v}\nabla)\mathbf{v} = \frac{\nabla p}{\rho} + \nu\nabla^2\mathbf{v} + \mathbf{g} \tag{2.27}$$

$$\frac{\partial \rho}{\partial t} = -\text{div}(p\mathbf{v}) \tag{2.28}$$

$$\frac{\partial c_i}{\partial t} = -\text{div}(\mathbf{N}_i) + R_i \tag{2.29}$$

$$-\nabla^2\Phi = \frac{F}{\epsilon}\sum_i z_i c_i \text{ or } \sum_i z_i c_i = 0. \tag{2.30}$$

Remark 2.2. *Equations (2.5) and (2.22) can be regarded as expressions of basic physical laws. They state that current is caused by the motion of charged particles and that individual species either are conserved or take part in homogeneous chemical reactions. However, the rate processes in the expression of the production rate and the flux density (Equation (2.10)) introduce uncertainties, e.g., the production rate involves chemical kinetics, for which rate expressions are neither predictable nor general, for details see[77]. In spite of these problems, the flux equation is recommended for general use because it is prevalent, both explicitly and implicitly, in the electrochemical literature and because it gives a good account of the physical processes involved without excessive complication. However, this is strictly valid only in dilute solution.*

It is assumed that the concentration of the species does not affect the velocity field. Hence, the fluid mechanics can be studied separately from the electrochemical analysis. To be more specific, in a first step the Navier-Stokes system (Equations (2.27) and (2.28)) is solved. The second step, makes use of the velocity field computed with the Navier-Stokes system to solve the electrochemical system. The first part is not discussed in this work. Information on solving incompressible steady state Navier-Stokes equation with algebraic multigrid (AMG) can be found in [103, and references therein].

2.4   Boundary Conditions

The Navier-Stokes system considered (Equations (2.27) and (2.28)) makes use of the boundary conditions

- $\mathbf{v} = \mathbf{v}_{\text{imposed}}$ or $p = p_{\text{inlet}}$ at the inlet,

- $p = p_{\text{outlet}}$ at the outlet, and

- $\mathbf{v}^T \mathbf{n}_b = 0$ and $\mathbf{v}_t = 0$ at walls,

where $\mathbf{n}_b$ is a vector normal to the boundary, $\mathbf{v}_{\text{imposed}}$ the imposed inflow velocity, and the p_{ref} reference pressure at the outlet.

Hence, the flux of a species normal to the boundary is (cf. Equation (2.10))

$$\overline{\mathbf{N}_i} := \mathbf{N}_i \frac{\mathbf{n}_b}{\|\mathbf{n}_b\|_2} = -F z_i u_i c_i \frac{\partial \Phi}{\partial \mathbf{n}_b} - D_i \frac{\partial c_i}{\partial \mathbf{n}_b} + 0. \qquad (2.31)$$

The boundaries of the electrochemical problem are composed of insulators and electrodes each with their own boundary conditions. We describe the boundary conditions for the insulators in Section 2.4.1 and the boundary conditions for the electrodes in Section 2.4.2.

2.4.1 Boundary Conditions on Insulators, Inlets and Outlets

On inlets, bulk values of the ion concentrations are imposed as Dirichlet conditions, such that the values obey the electroneutrality and homogeneous reaction equilibrium conditions. It is assumed that the inlet and outlet are in the far-field of the electrodes and, hence, $\nabla \Phi = 0$ holds. Furthermore, it is assumed that $\nabla c_i = 0$ holds at the outlet.

Since it is impossible for species and especially charges to leave the system through an insulating wall, the flux of a species perpendicular to an insulator is zero. We have

$$\overline{N_i} = F z_i u_i c_i \frac{\partial \Phi}{\partial \mathbf{n}_b} + D_i \frac{\partial c_i}{\partial \mathbf{n}_b} = 0. \tag{2.32}$$

In Section 2.1, we mentioned that a single driving force, i.e., the gradient of the electrochemical potential of a species, is appropriate for both diffusion and migration. In fact, ionic mobility and diffusion coefficient are related. This relationship is provided by the

Remark 2.3. *The Nernst-Einstein equation is strictly applicable only at infinite dilution. The quantities D_i and u_i in the Nernst-Einstein equation are not adequately defined at non-zero concentrations. For further information and for some remarks regarding the flow equation (which includes D_i, u_i) and concentrated electrolytes, see [77].*

Multiplying Equation (2.32) with z_i (only if $z_i \neq 0$) and $1/u_i$ ($u_i \neq 0$, since the fluid is not solid), summing over all species i (with $z_i \neq 0$), and using the Nernst-Einstein equation leads to

$$F \underbrace{\sum_{i, z_i \neq 0} (z_i^2 c_i)}_{a} \frac{\partial \Phi}{\partial \mathbf{n}_b} + \Re T \frac{\partial}{\partial \mathbf{n}_b} \underbrace{\left(\sum_{i, z_i \neq 0} z_i c_i \right)}_{=\sum_i z_i c_i = 0 \text{ electroneutrality eq.}} = 0. \tag{2.33}$$

$a \neq 0$ because of $c_i \geq 0$ and electroneutrality equation (2.2). Thus, the equation becomes

$$\frac{\partial \Phi}{\partial \mathbf{n}_b} = 0. \tag{2.34}$$

Substituting this into Equation (2.32), we get

$$\frac{\partial c_i}{\partial \mathbf{n}_b} = 0 \tag{2.35}$$

for each species i (with $z_i \neq 0$). If $z_i = 0$ we will directly get $\frac{\partial c_i}{\partial \mathbf{n}_b} = 0$ from Equation (2.32). Hence, the boundary conditions for insulating walls are described by Equation (2.34) and (2.35).

Remark 2.4. *We use the electroneutrality equation instead of Poisson's equation in order to derive the boundary conditions. This is justified because of the close adherence of electrolytic solutions to the condition of electroneutrality, see [77] for details.*

Remark 2.5. *There are always at least two $z_i \neq 0$, since without this fact it would not be an electroneutral electrochemical system.*

2.4.2   Boundary Conditions on Electrodes

Usually, not all species react at the electrodes. This results in two separate boundary conditions describing the heterogeneous reactions - one for reacting species and one for non-reacting species. We use the following nomenclature in order to distinguish between reacting and non-reacting species. The subscript i denotes a general species, l is used to denote a reacting species and m is used to declare a non-reacting species. Bold letters I, L, M denote the number of species, respectively. Therefore, we have $I = L + M$. Additionally, we declare the K reactions taking place at the electrodes with the subscript k.

Boundary Conditions for Reacting Ions

When an electrode is in contact with the solution, a new arrangement of solvent dipoles, species in the solution, and electrons in the electrodes is obtained at the contact surface. The homogeneous and isotropic character of the electrode and the solution is disturbed. Equal and opposite charge concentrations arise on each side of the contact surface and, consequently, an electromagnetical field is built up. This charge separation, called the *electronic double layer*, develops a potential difference across the interface. The value of this potential difference depends on the nature of electrode and solution. The double layer is electrically equivalent with a parallel-plate capacitor. A detailed description of the double layer can be found in the literature [77].

The absolute value of the potential difference across an electrode-electrolyte interface cannot be measured since each attempt to do that would introduce a new electrode-electrolyte interface. Therefore, a reference electrode known as *normal hydrogen electrode* (NHE) is used to make relative measurements possible. The potential difference at equilibrium measured with respect to a NHE is called equilibrium potential or *Nernst potential* $E_{k,\mathcal{R}_k}$ of reaction k, where $\mathcal{R}_k := \{$species $l\ |l$ is involved in reaction $k\}$. Therefore, we can write

$$V = \tilde{\phi} + E_{k,\mathcal{R}_k}, \tag{2.36}$$

with $\tilde{\phi}$ the potential of the solution adjacent to the electrode, and V the electrode potential. The equilibrium at the electrode-electrolyte interface is to be considered as a dynamic situation for each species $l \in \mathcal{R}_k$. Indeed, there is a continuous exchange of charges between the two phases, but without net current. Therefore, at equilibrium oxidation and reduction reactions occur simultaneously and with the same rate.

A driving force is required in order to obtain an electrochemical reaction k on an electrode at a given rate involving species $l \in \mathcal{R}_k$. This driving force is the potential difference which is different from the Nernst potential. It is called *overpotential* for charge transfer η_k and occurs between the electrode potential V and the potential in the solution $\tilde{\phi}$ which is adjacent to the electrode. Therefore, we have for each reaction k

$$V - \tilde{\phi} = E = \eta_k(\bar{\mathrm{j}}_k, c_{\mathcal{R}_k}) + E_{k,\mathcal{R}_k}, \tag{2.37}$$

where E denotes the galvanic potential, $\bar{\mathrm{j}}_k$ the current density which is normal to the electrode surface caused by reaction k, and $c_{\mathcal{R}_k}$ the concentrations of all species $l \in \mathcal{R}_k$ involved in reaction k adjacent to the electrode.

When several species react on an electrode, the total current density going through the electrode can be seen either as the sum of the current densities coming from each reacting species, or as the sum of the reacting currents. Hence, the set of equations describing the boundary conditions is given by

$$\bar{\mathrm{j}}_{tot} = \sum_k \bar{\mathrm{j}}_k = \sum_l \bar{\mathrm{j}}_l$$

$$\tilde{\phi} = V - \eta_k(\bar{\mathrm{j}}_k, c_{\mathcal{R}_k}) - E_{k,\mathcal{R}_k}. \tag{2.38}$$

The contribution of each reaction to the total current density is such that the galvanic potential is constant.

The current density of each reaction k cannot be calculated directly from the total current density normal to the electrode, because one species $l \in \mathcal{R}_k$ can contribute to different reactions. However, we can write

$$\bar{\mathbf{j}}_l = z_l F \overline{\mathbf{N}}_l = \sum_k s_{lk} \bar{\mathbf{j}}_k, \tag{2.39}$$

$$\text{with} \qquad \sum_l s_{lk} = 1 \tag{2.40}$$

$$\text{and} \quad \overline{\mathbf{N}}_l = F z_l u_l c_l \frac{\partial \Phi}{\partial \mathbf{n}_b} + D_l \frac{\partial c_l}{\partial \mathbf{n}_b}, \tag{2.41}$$

where $\bar{\mathbf{j}}_l$ and $\overline{\mathbf{N}}_l$ are normal to the electrode. s_{lk} are defined by stoichiometry of each reaction k involving species $l \in \mathcal{R}_k$. The remaining unknown $\bar{\mathbf{j}}_k$ is also normal to the electrode and defined by the non-linear Butler-Volmer equation. The Butler-Volmer equation describes how the electrical current on an electrode depends on the potential difference across the electrode/electrolyte interface and the local concentration of the reacting ions. Details on the Butler -Volmer equations can be found, e.g., in [32, 75, 77].

Boundary Conditions for Non-Reacting Ions

For each non-reacting species m the following equation holds

$$\bar{\mathbf{j}}_m = z_m F \overline{\mathbf{N}}_m = 0. \tag{2.42}$$

Hence, the boundary condition becomes

$$\overline{\mathbf{N}}_m = 0 \;\; (\text{if } z_m \neq 0), \tag{2.43}$$

with $\bar{\mathbf{j}}_m$ and $\overline{\mathbf{N}}_m$ normal to the electrode. Using Equation (2.31) and applying Nernst-Einstein's equation (??) we have

$$\frac{\partial c_m}{\partial \mathbf{n}_b} = -\frac{F z_m c_m}{RT} \frac{\partial \Phi}{\partial \mathbf{n}_b} \;\; (\text{ if } z_m \neq 0)$$

$$\text{or } \overline{\mathbf{N}}_m = -D_m \frac{\partial c_m}{\partial \mathbf{n}_b} \;\; (\text{ if } z_m = 0), \tag{2.44}$$

where $\mathbf{n}_b$ denotes the normal vector of the electrode.

2.5   Existence and Uniqueness of Solutions

The existence and uniqueness of a solution for the general multidimensional MITReM equations has not been proved so far. However, studies show that the simulation results do describe the reality sufficiently well. Furthermore, several proves for the 1D case without convection and for the case of binary electrolytes exist.

Sokirko and Bark [95] consider a steady-state one-dimensional migration-diffusion model. The electrolyte in the system investigated is made up of three ionic species. Only one of the species takes part in the electrode reactions. An exact expression for the polarization curve, which describes voltage and current density, is given in implicit form. In the special case of a binary electrolyte, an exact explicit expression is given for the polarization curve.

In Bortels, van den Bossche, Deconinck, Vandeputte, and Hubin [10] a generalization of Sokirko and Bark [95] is given. Bortels et al. obtain an analytical solution for the one-dimensional steady-state transport of ions in an electrolyte towards a planar electrode. Additionally, an implicit form of the solution is given for more than one electroactive species and any number of non-reacting species.

The full (migration - diffusion - convection - reaction) MITReM strongly resembles the drift-diffusion(-convection-reaction) systems to be solved in semiconductor physics when it comes to a simulation of the electrostatic behavior of semiconductor devices such as diodes or transistors ("device simulation"). There, theoretical work on existence and (local) uniqueness of solutions has been done by several authors, see for instance [38, 70, 71, 94] and references given therein. We believe that the existence and uniqueness for the MITReM can be shown with similar techniques.

In the following, we briefly describe the existence and uniqueness for the most simple case of binary electrolytes.

The Binary Electrolyte

A binary electrolyte consists only of two ions, which do not show homogeneous reactions. Hence, the system to be solved is

$$\frac{\partial c_i}{\partial t} = F z_i u_i \nabla c_i \nabla \Phi + F z_i u_i c_i \Delta \Phi + D_i \Delta c_i - \mathbf{v} \nabla c_i \qquad (2.45)$$

$$0 = \sum_i z_i c_i, \qquad (2.46)$$

with $i = 1, 2$. In this special case, we can easily compute an analytical solution if the boundary and initial conditions are properly set. In order to show this and to demonstrate the chemical properties, we transform the original system. Furthermore, the transformed system is far more easy to handle in order to obtain an analytical solution. For simplicity, we use the system formulation with the electroneutrality equation instead of the Poisson equation (cf. Section 2.1).

In order to create the equivalent system we substitute the concentration c_1,

$$c_1 = \frac{z_2}{z_1} c_2, \tag{2.47}$$

and obtain

$$\nabla c_2 \nabla \Phi \; + \; c_2 \Delta \Phi = \frac{1}{F z_1 u_1} \left[\frac{\partial c_2}{\partial t} - D_1 \Delta c_2 + \mathbf{v} \nabla c_2 \right] \tag{2.48}$$

$$\frac{\partial c_2}{\partial t} \; = \; F z_2 u_2 \nabla c_2 \nabla \Phi + F z_2 u_2 c_2 \Delta \Phi + D_2 \Delta c_2 - \mathbf{v} \nabla c_2. \tag{2.49}$$

Substituting $\nabla c_2 \nabla \Phi + c_2 \Delta \Phi$ in Equation (2.49) leads to

$$\frac{\partial c_2}{\partial t} + \mathbf{v} \nabla c_2 \; = \; \frac{z_2 u_2}{z_1 u_1} \left[\frac{\partial c_2}{\partial t} - D_1 \Delta c_2 + \mathbf{v} \nabla c_2 \right] + D_2 \Delta c_2. \tag{2.50}$$

Define constants K and D to be

$$K \; = \; 1 - \frac{z_2 u_2}{z_1 u_1}, \tag{2.51}$$

$$D \; = \; D_2 - D_1 \frac{z_2 u_2}{z_1 u_1}. \tag{2.52}$$

Since $K \neq 0$, the new system is given by

$$\nabla c_2 \nabla \Phi + c_2 \Delta \Phi \; = \; \frac{1}{F z_1 u_1} \left[\frac{\partial c_2}{\partial t} - D_1 \Delta c_2 + \mathbf{v} \nabla c_2 \right] \tag{2.53}$$

$$\frac{\partial c_2}{\partial t} + \mathbf{v} \nabla c_2 \; = \; \frac{D}{K} \Delta c_2 \tag{2.54}$$

$$c_1 \; = \; \frac{z_2}{z_1} c_2. \tag{2.55}$$

The second equation is independent of Φ and c_1. Hence, an analytical solution for the system considered can be computed if the boundary and initial conditions are properly set.

2.6   Discretization and Linearization

Finite differences, finite volumes and finite elements are established methods to discretize PDE systems. However, in recent years, several hybrid discretization methods have been developed which combine ideas of these three methods in order to gain greater flexibility, accuracy or robustness.

One of these hybrid methods is the residual-distribution scheme. The residual-distribution scheme, also known as fluctuation-splitting or cell-vertex scheme, was developed at the end of the nineties of the last century and provides an alternative to classical upwind finite volume schemes. The residual-distribution scheme is based on a piecewise linear representation of the flow variables on simplices similar to finite elements. It provides a lower cross-diffusion than their finite volume counterparts, see [83, 114, and references therein].

The considered continuous MITReM PDE system writes

$$
\frac{\partial c_i}{\partial t} = \boxed{F z_i u_i \nabla c_i \nabla \Phi} + \boxed{F z_i u_i c_i \Delta \Phi} + D_i \Delta c_i - \mathbf{v} \nabla c_i + \boxed{R_i} \quad (2.56)
$$

$$
\nabla^2 \Phi = -\frac{F}{\epsilon} \sum_i z_i c_i. \quad (2.57)
$$

The non-linear terms are marked with boxes. The employed simulation software makes use of a combined residual-distribution finite element discretization on a triangular grid. The mathematical justification for the validity of the mixed discretization can be done analogously to [35]. In order to linearize the system, Newton's method is employed.

Newton's method is a root-finding algorithm which is frequently used to linearize non-linear PDEs. It generally gives a quadratic convergence near the root. However, the method can be unstable near a horizontal asymptote or a local extremum which can result in worse convergence or even divergence.

The diffusion, migration and reaction terms are discretized via finite elements using linear shape functions, see Section 2.6.2. Using this technique in the case of the convective term would lead to massive problems due to checkerboard instabilities. Hence, the convective term is discretized using the N-Scheme of the residual-distribution method which provides stability, positivity and robustness. We describe the details of the method in Section 2.6.3.

The migration term does also have a convective character. In Chapter 7, we demonstrate that the checkerboard instabilities occur for the migration near the electrodes. The instabilities are caused by a violated Péclet condition. We give a brief description of the phenomenon for the case of the scalar convection-diffusion equation in Section 2.6.1.

2.6.1   The Péclet Number

We briefly describe checkerboard instabilities, which may appear in the context of convection-dominated applications. The existence of such instabilities depends on the discretization used, see, e.g., [50, 106] for more details on this topic.

Considering the 1D convection-diffusion equation

$$-Du_{xx} + u_x = f \text{ in } \Omega, \tag{2.58}$$

where $\Omega = (0,1)$, $D \in \mathbb{R}$, $D > 0$ is the diffusion constant, f a given scalar function, and u the scalar unknown function to be solved for. D is assumed to be constant. Furthermore, we use Dirichlet boundary conditions

$$u(0) = 0, \quad u(1) = 1. \tag{2.59}$$

The solution of the problem is given by

$$u(x) = \frac{1 - e^{\frac{x}{D}}}{1 - e^{\frac{1}{D}}}. \tag{2.60}$$

Note that the solution has a sharp boundary near $x = 1$, see Figure 2.2.

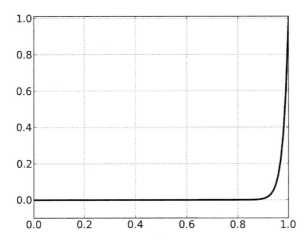

Figure 2.2: $u(x) = \dfrac{1 - e^{\frac{x}{1/50}}}{1 - e^{\frac{1}{1/50}}}$ (D=1/50).

Discretizing the problem with second order central finite differences on a regular grid with mesh size h, we obtain the discrete solution

$$u_h(x_k) = \frac{1 - \left(\frac{2D+h}{2D-h}\right)^k}{1 - \left(\frac{2D+h}{2D-h}\right)^N}, \tag{2.61}$$

where $N = 1/h$ and $x_k = kh$ $(k = 0, ..., N)$. Obviously, the solution tends to high oscillations for $h/D > 2$, see Figure 2.3.

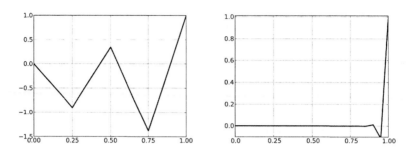

Figure 2.3: Discrete solutions for $h = 1/4$ and $h = 1/20$, $D = 1/50$.

If we use the second order central finite difference operator in 1D

$$L_h = \frac{1}{h^2} [-h/2 - D \ \ 2D \ \ h/2 - D]_h, \qquad (2.62)$$

it is obvious that any grid function,

$$f(x_k) = \begin{cases} c_1 & \text{if } k \text{ is odd} \\ c_2 & \text{if } k \text{ is even} \end{cases}, \qquad (2.63)$$

will lie in the kernel of the discrete operator for $D = 0$. This is also called checkerboard-type instability. A similar behavior is observed for higher dimensional problems, too.

The mesh Péclet number for a convection-diffusion problem on a regular grid with mesh size h is defined by

$$\overline{\text{Pe}} := \frac{h}{D} \|\mathbf{v}\|_\infty, \qquad (2.64)$$

where D again is the diffusion coefficient and $\mathbf{v} \in \mathbb{R}^n$ the velocity vector.

With respect to the above example $\overline{\text{Pe}}$ has to be lower or equal 2 in order to avoid oscillations. This restriction is called **Péclet condition**.

There is no unique way to define a discrete Péclet number in the context of finite element discretizations. Various definitions are used in literature. Two common definitions of the element Péclet number Pe_{e_k} for a finite element e_k are

- $\text{Pe}_{e_k} = \frac{v_{e_k} \max_{i=1,...,M}(l_i)}{D}$ where M is the number of edges of the element, l_i the length of edge i of e_k, and v_{e_k} the average element velocity, and

- $\text{Pe}_{e_k} = \frac{\sum_i \|\mathbf{v}_i \nabla N_i\|}{D}$ where N_i is the shape function $i = 1, ..., M$ and $\mathbf{v}_i$ is the velocity in i.

The first method is commonly used for theoretical studies, whereas the latter often turns out to be meaningful in practice, for instance, when considering highly irregular grids with stretched elements [102].

Checkerboard instabilities which are caused by a high Péclet number can be avoided when choosing a fine enough discretization, introducing artificial viscosity/dissipation or using upwind (biased) discretizations. We briefly describe common stabilization techniques for the finite element method in Section 2.6.2.

2.6.2   Finite Element Method

The employed finite element method makes use of the Ritz-Galerkin approach using linear shape functions, details on the finite element method can be found, e.g., in [12, 28, 96].

It is well-known that the finite element discretization can cause stability problems when used for convection dominated applications. These problems lead to non-physical oscillations which pollute the numerical solution. There exist several techniques in order to stabilize the finite element method in this case, see, e.g., [72, 88, and references therein]. Well-known stable finite element formulations are, e.g., streamline-upwind Petrov-Galerkin method see, e.g., [51, 52, 55], residual-free bubble method, see, e.g., [36], and the discontinuous finite element method, see, e.g., [30].

Another possibility is to use finite element methods with special elements to stabilize the discrete system, like Taylor-Hood-Elements or the more accurate, but also more work intensive, Crouzeix-Raviart-Elements [23, 59].

Besides the stabilizations mentioned above, there are other stabilization possibilities which are not based on standard finite element methods and instead try to use another discretization which has inherited stabilization properties, e.g., staggered grids [40, 47, 64, 78] or hybrid discretizations like the residual-distribution scheme which is used in the simulator for discretizing the convection.

The migration term dominated near the electrodes. Furthermore, the grid in this area of the domain is only highly refined in one direction. Since the migration also has a convective character, the Péclet condition is violated near the electrodes. However, this violation does not affect the overall solution of the system significantly. That is, the results do still represent the reality sufficiently well in the considered applications. Hence, there is no need to change the discretization for the migration.

2.6.3   Residual-Distribution Scheme

The residual-distribution scheme computes residuals over so-called primary elements and distributes them to the nodes, which then collect the residuals of the neighboring elements to compute the new values at the nodes. Each node thereby accumulates a residual contribution from each of the primary elements to which it belongs. The sum is then used to update the nodal value in either an explicit or implicit manner.

Residual-distribution schemes offer a number of advantages over the traditional finite-volume approaches. One big advantage of this method is its good accuracy even on irregular grids. Furthermore, the method allows the construction of highly compact stencils whose accuracy rivals that of traditional finite-volume methods requiring a much larger support because of the natural way in which they enable truly multidimensional or streamline upwinding.

Dependent on the underlying problem or/and the properties required, there are several schemes which can be used to distribute the residuals to the nodes, e.g., LDA [97] or the simplified Lax - Wendroff scheme for Navier-Stokes problems [4, 103].

For transport equations, the N-Scheme is used because it provides positivity, however, at the cost of a lower accuracy. We use the N-Scheme to discretize the convective term of the MITReM. The positivity of the scheme ensures that the concentration of the ions cannot become negative. We should note that the N-Scheme is only of first order accuracy when computing on general irregular grids. However, in the case of structured and aligned grids, it provides nearly second order accuracy [114].

The N-Scheme of the residual-distribution method is the analogon of the first-order multidimensional upwind method in the case of finite differences although it has got a better accuracy than the common upwind-scheme [82].

Chapter 3

Algebraic Multigrid (AMG) for Convection-Dominated Flow Problems

Multigrid methods are a methodology to create linear solvers. They represent a whole class of different approaches. All multigrid approaches have in common a hierarchy of levels/grids and essentially combine two ingredients - **smoothing** and **coarse-grid correction**.

Given a linear system and an approximate solution to it, the error can be classified into two parts. Only one part of the error is reduced effectively by the smoother. If the coarse-grid correction process is able to reduce the "remaining" part of the error effectively, the result will be an optimally scaling solution approach.

The algebraic multigrid (AMG) technology is a basis for a whole class of different solution techniques. The most common techniques are aggregative AMG and classical AMG. In this thesis, we consider the point-based algebraic multigrid (PAMG) technique to solve the MITReM. PAMG is a generalization of classical AMG for strongly coupled systems of PDEs.

An important property of AMG methods is that they can be applied directly to structured and unstructured grids because they make only use of the linear system itself and do not need any additional information. This property makes them very popular for industrial applications since they can easily be integrated into existing software. However, if only considering the linear system itself, the optimality of such an approach is, in many cases, not guaranteed. This is especially true for convection-dominated problems like the MITReM, which arise in many industrially relevant applications.

Using state-of-the-art PAMG techniques to create linear solvers and applying these solvers to the MITReM does not lead to an effective solution approach. The reason being that physical properties of the PDE system are not represented well in the standard PAMG framework. In this chapter, we introduce the standard PAMG framework and techniques commonly used to solve convection-dominated problems with AMG methods. The new physics-aware PAMG techniques which lead to a robust and efficient approach are introduced in the following chapters.

We briefly introduce the AMG technology in Section 3.1. Then, we introduce state-of-the-art PAMG in Section 3.2. Finally, we give a survey on common techniques to solve convection-dominated problems with AMG methods in Section 3.3.

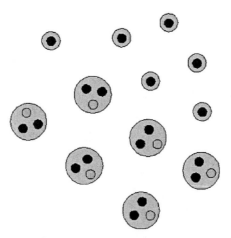

Figure 3.1: Points, variables and unknowns [29].

Notation 3.1. *We distinguish between points, unknowns and variables, cf. Figure 3.1. Points are positioned where the values of the solution of a linear system considered are computed. In the figure, the points are the big gray bubbles. Points are usually identified with the grid points. The variables are the entries of the solution vector, in the figure these are the small colored bubbles. Finally, unknowns represent physical functions to be solved for; they are, e.g., identified with velocity components or pressure. In the figure, variables belonging to the same unknown have got the same color.*

3.1   AMG

We first introduce the AMG technology in Section 3.1.1. Then, we briefly summarize common AMG variants in Section 3.1.2.

3.1.1   The AMG Technology

We briefly describe the basic components and principles of the AMG technology. A detailed description can be found, e.g., in [100].

The AMG technology consists of four main components

- smoothing,

- coarse level selection,

- restriction, and

- interpolation.

It proceeds in two phases. In the first phase, called **setup**, the coarse levels are selected, and the interpolation and restriction are set up accordingly. Furthermore, the smoother may be initialized if necessary, e.g., computing the LU-factorization for each level if using ILU-type smoothing. In the second phase, called **cycling**, the components are applied iteratively to compute the solution of the linear system considered.

A two-level AMG cycle proceeds as follows

1. perform relaxations on the fine level problem until the error is algebraically smooth,

2. compute the residual and transfer it to the coarser level,

3. solve the coarse level defect equation to obtain a correction,

4. interpolate the correction to the fine level and update the fine level approximation, and

5. perform final relaxation steps.

Points 2-4 build up the coarse-grid correction. The two-level cycle is applied recursively to obtain a complete AMG cycle.

In order to obtain a robust and efficient AMG solver the coarsening and interpolation have to be defined such that the overall coarse-grid correction supports the smoothing process chosen. Each type of AMG has a different philosophy how to satisfy this in the best way. Common AMG variants are summarized in the next section.

In the case of systems of partial differential equations or involved scalar problems, often, matrices which cannot be solved by stand-alone AMG methods arise. The efficiency and robustness of AMG can be improved by using it as a preconditioning for Krylov methods like GMRES, BICGSTAB or CG, see, e.g., [91] for details on Krylov methods.

Hence, we say that Krylov methods accelerate multigrid methods. We show that accelerated PAMG leads to a much better convergence behavior and saves run time compared to stand-alone PAMG in the case of the MITReM, c.f. Section 7.3.1.

3.1.2   AMG Variants

Classical AMG has been developed in the early eighties for solving systems $Av = b$ with weakly diagonally dominant Stieltjes matrices [16, 17, 98]. The theoretical basis has been given in Brandt [13]. The first realization of classical AMG is described in Ruge and Stüben [89].

In the nineties of the last century, classical AMG approaches have substantially been enhanced because of growing problem sizes in industrial applications and a higher complexity of the arising problems. This development resulted in more robust interpolation formulas, namely standard interpolation and multi pass interpolation, and aggressive coarsening strategies, reducing memory requirements considerably, cf. [60, 99].

Aggregation-based multigrid has been developed in parallel to the improvement of classical AMG methods, see, e.g., [6, 11, 67, 110, 111]. The main difference to classical AMG is that aggregative AMG combines variables to aggregates to form the coarser level, where classical AMG selects variables to be in the coarser level. The simplest interpolation in the context of aggregative AMG is piecewise constant interpolation, where each fine level variable in an aggregate receives the same interpolation formula. The simplicity of this method leads to a large acceptance in the multigrid community.

Aggregative multigrid with piecewise constant interpolation can be regarded to be a special case of classical AMG. Namely, it has the aim of keeping coarsening and interpolation as simple as possible. This results in a very cheap method regarding computational work and memory. There exist many variants of aggregative AMG the most popular one is smoothed aggregation multigrid (SAAMG). SAAMG uses an a-posteriori improvement of the interpolation by employing a smoothing process before the Galerkin operator is computed. It is characterized by a better run time and generality than the standard aggregative multigrid method, which only makes use of piecewise constant interpolation. A theoretical analysis of the convergence rate of aggregative methods can be found, e.g., in Napov and Notay [73], Vaněk et al. [111].

Besides the classical and aggregative algebraic multigrid approaches, there exist several other AMG approaches. Many of these approaches make use of additional information. They are based on element preconditioning [44], element interpolation [21, 27, 49, 56], and energy minimization [25, 68, 113].

Furthermore, adaptive AMG variants have become an active research field in the last ten years. The basic idea, compatible relaxation (CR), has been introduced in [14], where the coarse level variables for classical AMG are determined by analyzing the behavior of the relaxation. This idea has been transfered to other AMG-type approaches [19, 20, 22, 65]. The adaptive concept has been extended to interpolation in [15]. This approach is called bootstrap AMG [69]. Combining CR and bootstrap approaches will become more and more popular to solve problems hardly solvable with the classical approaches [18]. However, the adaptivity has the drawback of a computational overhead which only pays off if the standard approaches cannot handle the problems considered.

3.2 PAMG

PAMG is a generalization of classical AMG for strongly coupled systems of PDEs. It has been developed in [90], and extended and successfully applied to drift-diffusion systems from device simulation in [29]. The basic idea of this approach is to solve all physical unknowns of the system simultaneously. Besides the linear system considered, PAMG makes use of the variable-unknown and variable-point mappings.

PAMG uses couplings between points to define coarsening and interpolation. These point-couplings are represented by point-coupling matrices. The couplings between the unknowns of two points are replaced by a scalar value which describes the point-coupling. All scalar values are represented in the primary matrix. On the basis of this primary matrix the C-points, which exist at the coarser level, and the F-points, which only exist at the finer level, are set up. Since PAMG works with point couplings, this results in the same hierarchy for all unknowns.

We briefly describe the main components of PAMG, i.e., smoothing (Section 3.2.1), coarse level selection and restriction (Section 3.2.2), and interpolation (Section 3.2.3).

3.2.1 Smoothing

The main area of application for PAMG are systems with strong unknown cross-couplings. In this case, it is often most appropriate to perform a point-oriented smoothing, i.e., treating variables belonging to the same point simultaneously. This should especially be true for applications where the decisive unknown cross-couplings are located mostly on the block-diagonal of the matrix (if considering a point-wise ordering), like in the case of the MITReM.

3.2.2 Primary Matrices and Point-Coarsening

In PAMG, the coarsening is done with the help of a "primary" matrix. There are several possibilities to choose this primary matrix. The aim is to set up a matrix which leads to a coarsening in the "direction" of smooth error.

Subsequently, we assume the matrix entries in the original matrix to be ordered point-wise. Hence, it can be written in the form

$$A = \begin{pmatrix} A_{(1,1)} & A_{(1,2)} & \cdots & A_{(1,n-1)} & A_{(1,n)} \\ A_{(2,1)} & A_{(2,2)} & \cdots & A_{(2,n-1)} & A_{(2,n)} \\ \vdots & \vdots & \ddots & \vdots & \vdots \\ A_{(n-1,1)} & A_{(n-1,2)} & \cdots & A_{(n-1,n-1)} & A_{(n-1,n)} \\ A_{(n,1)} & A_{(n,2)} & \cdots & A_{(n,n-1)} & A_{(n,n)} \end{pmatrix}, \qquad (3.1)$$

where n denotes the number of points and $A_{(k,l)}$ is the point-coupling matrix of dimension "number of unknowns in point k" $\times$ "number of unknowns in point l". Each of the point-coupling matrices shall be assigned a scalar value, which represents the coupling strength between points k and l. The scalar values set up the primary matrix of dimension $n \times n$ which builds the basis for the coarsening process.

Natural possibilities for the choice of the primary matrix are

- define the primary matrix on the basis of a scalar auxiliary problem,

- choose an primary unknown which has to be defined at all points, and

- choose a primary matrix dependent on norms of the coupling blocks $\boldsymbol{A}_{(k,l)}$.

When defining the primary matrix on the basis of a scalar auxiliary problem additional information has to be given. If we consider, for example, anisotropies in a given problem which arise because of non-uniform mesh spacings, a suitable primary matrix might be given by a Laplace operator. If coordinates are available, this might be achieved by a coordinates-based primary matrix. However, if coordinates are not available an approximate discrete Laplacian might be used as the primary matrix. This possibility also helps if an appropriate auxiliary problem cannot be defined, but underlying physics are well known.

When an unknown is available at all points, it can be chosen as a primary unknown. In this case, only the couplings of the unknown to itself represent the primary matrix.

In the norm-based approach, the matrix norm of $\boldsymbol{A}_{(k,l)}$ is computed. From the practical point of view, there are two main strategies for setting up a primary matrix $\mathbf{P} = (p_{ij})$ with the help of norms

- $p_{kl} = -\|\boldsymbol{A}_{(k,l)}\|$ $(k \neq l)$ and $p_{kk} = \|\boldsymbol{A}_{(k,k)}\|$, and

- $p_{kl} = -\|\boldsymbol{A}_{(k,l)}\|$ $(k \neq l)$ and $p_{kk} = -\sum_{k \neq l} p_{kl}$.

Both approaches can be used with various norms. Commonly, the maximum-norm is used because it is computationally cheap.

It is obvious that the signs of the couplings between points are ignored. Hence, the information of positive or indefinite point-coupling matrices gets lost. This is only for practical reasons, it saves computational work and it also is not clear if the original matrix is treated the right way even if we consider the signs, see [29] for further explanations.

Note, in order to assign the sparsity of the original matrix to the primary matrix, it is reasonable to assign a non-zero value only if the two points k and l are coupled.

After setting up, the primary matrix is treated as if it were given into classical scalar AMG. That is, the algorithm creates the coarsening strategy for the primary matrix which is then transfered to the original matrix to be solved for.

3.2.3   Interpolation

The following interpolation possibilities for PAMG are commonly used

- block interpolation (B-interpolation),

- multiple-unknown interpolation (MU-interpolation), and

- single-unknown interpolation (SU-interpolation).

The B-interpolation takes couplings between different unknowns into account and computes the interpolation formulas block-wise. This approach seems to be natural if the primary matrix is defined on norms. In practice, this approach needs a lot of computational work and is technically cumbersome (see [29]).

The MU-interpolation is computed variable-wise and separately for each unknown. The fine level variables are only interpolated from variables of the same unknown. It may be used in cases where only the coarsening should be computed point-wise.

The SU-interpolation also uses variable-wise interpolation formulas like the MU-interpolation, but the formulas are the "same" for the variables belonging to the same point. Furthermore, the fine level variables are interpolated from variables of the same unknown. The interpolation for the set of points is based on the primary matrix. Namely, the interpolation is computed as in the case of classical scalar AMG for the primary matrix, and then the interpolation scheme is transfered to the variables belonging to each point similar to the coarsening. Note that the primary matrix for interpolation and the one for coarsening do not have to be the same. However, this is often the best choice in the case of strong unknown cross-couplings.

3.3 AMG for Convection-Dominated Flow Problems

We summarize results from literature which deal with the application of AMG to convection-dominant flow problems. Since these kind of problems frequently occur in industrial applications, there exist many publications in this area. Therefore, we restrict the discussion to the most important aspects which have to be considered for the construction of an appropriate AMG method.

In convection-dominated applications, usually very robust smoothers, e.g., ILU-type or smoothers which capture the flow are used. If the smoother has to capture the direction of flow, the ordering of variables has to be chosen accordingly. Techniques which deal with this aspect are discussed in Chapter 4, where we develop the reordering framework.

However, problems may also occur even if smoothing works well because the use of coarser levels introduces a new source of problems. To be more specific, even if the initial problem satisfies the restrictions introduced by the discretization, e.g., Péclet condition (see Section 2.6.1 for details), one or more coarser grids may violate them.

Considering incompressible Navier-Stokes and Euler equations, one strategy is to circumvent problems which might be caused by checkerboard instabilities by decoupling the system (i.e. SIMPLE-approach for Navier-Stokes, Griebel, Neunhoeffer, and Regler [41] or Roberts, Sidilkover, and Swanson [87] for Euler), or by treating the convection explicitly (see for instance Larson, Synder, Abeele, and Clees [61]). In both cases the linear systems to be solved with AMG are elliptic. If the convection is treated explicitly, it often is beneficial to use the pressure-unknown couplings to compute the C/F-splitting (see for instance Raw [85], Weiss, Maruszewski, and Smith [116]).

In the case of geometric multigrid, problems may arise on coarse levels. Various techniques for dealing with these problems caused by checkerboard-type instabilities on coarser levels have been proposed. One possibility is to improve the coarse-grid correction with help of coarse-grid redistribution [24], i.e., to define more coarse level points in critical areas. Another technique is to add artificial dissipation on the coarse levels [33]. Johannsen [54] uses a Frobenius-complement Gauss-Seidel smoother in conjunction with a scaling of the Péclet number on coarse levels.

In the context of AMG methods, the phenomenon of increasing Péclet numbers on coarser levels arises, too. However, this is not true for all types of AMG. Considering the classical AMG approach applied to ordinary scalar convection-diffusion equations, the convergence factors are nearly independent of the grid resolution. This is true even if we consider convection-diffusion with closed characteristics, cf. Stüben [99]. Note that this is only valid if the Péclet condition is fulfilled on the fine level. Otherwise, the Péclet number is likely to increase on coarser levels since the matrix is not of M-matrix type anymore. We investigate this in Section 5.1.

In Okusanya, Darmofal, and Peraire [80] the authors use a stabilized finite element discretization for Navier-Stokes equation. The linear system is solved by agglomerative AMG. Since the dissipation on coarse grids relatively looses in size, additional dissipation is added on the coarse grids. Note, for this approach the discretization has to be split and the information has to be available for the AMG method.

Applying aggregative AMG to convection-dominated flow, either artificial viscosity is introduced on the coarser levels de Zeeuw and van Asselt [33], or interpolation and restriction operators are chosen differently of each other, i.e., the restriction is not the transposed of the interpolation Bank, Wan, and Qu [3], Guillard and Vaněk [42], Webster [115].

Besides improving the coarse-grid correction, the smoother can be improved by various techniques. In the case of higher order discretizations, a smoother based on the first order discretization might be sufficient, see Olshanskii and Reusken [81]. In the context of compressible Euler systems Naumovich, Förster, and Dwight [74] propose a defect-correction smoothing approach. AMG methods solve linear problems derived from first order discretizations of the compressible Euler equations well. Hence, linear problems resulting from second order discretizations which are of practical interest are solved by applying AMG to the first order system within a defect-correction iteration. Note that this idea has also been used before in the context of geometric multigrid [48].

All findings suggest that there exists no optimality for multigrid methods using higher order discretization with respect to increasing Péclet numbers.

Remark 3.2. *Unless stated otherwise, the terms "AMG" stands for classical AMG throughout this thesis. Furthermore, "VAMG" stand for classical scalar AMG throughout this thesis.*

Chapter 4

The Reordering Framework

Considering convection-dominated problems like the MITReM, the smoother of the AMG method has not only to smoothen the error, but also to reduce the "convective" part of the error. Applying the smoother employed as a stand alone one-level iterative solver to convection-dominated flow problems, a suitable ordering of the variables is often a prerequisite for fast converging solving approach. Hence, the ordering of the variables also has a significant influence on the convergence factor of a multigrid method.

In many simulation codes, the initial ordering of variables is induced by the grid-generation tool. Whether or not this initial ordering is beneficial for the linear solver employed is not usually investigated. The MITReM simulator considered makes use of a reverse Cuthill McKee (RCM) ordering of the grid points which reduces the bandwidth of the arising linear systems. Hence, this ordering is advantageous for ILU-type solvers, like the ILU-GMRES method employed which we want to replace with PAMG.

However, if we use an AMG approach with Gauss-Seidel type smoothing an ordering in the **direction of the dominant flow** (convection and/or migration) will drastically enhance the efficiency and robustness of the approach compared to a RCM ordering of variables.

The ordering of variables is usually computed with the help of the original matrix. Sometimes, information from the discretization is used. However, when applying state-of-the art reordering techniques to the MITReM, the resulting orderings are often very unstructured. This has several reasons, first of all, these reordering techniques are usually defined for scalar PDEs. Second, the MITReM describes two different directions of movement (migration and convection) which has not been considered so far. And third, it has not been investigated in which way explicitly given physical information like vector fields can be employed to create such orderings.

We develop a **reordering framework** which can be used to reorder linear systems derived from scalar as well as systems of PDEs. It makes use of a modular concept, so, it can easily be extended and equipped with new components. Especially, it is able to use additional (physical) information within the reordering procedure. The aim of this concept is to make the framework applicable in all simulation environments where a reordering helps to improve the efficiency of the linear solver employed - independent of the application and type of the linear solver.

In order to create a suited permutation for a given linear system, the reordering framework proceeds as follows

- choose a basic matrix,

- reduce the basic matrix (optional), and

- apply a (weighted) sorting algorithm.

In the first step, we choose a basic matrix. The basic matrix can either be the original linear system, or some auxiliary matrix. In the case of strongly coupled systems of PDEs, for example, its dimension is the number of points instead of the number of variables. In this case, the couplings of one unknown to itself can be used as the basic matrix.

In the second step, the basic matrix may be reduced, so that only "important" couplings remain in the matrix. This step is optional and might not be necessary to achieve a suitable ordering. In order to identify "important" couplings, external information such as vector fields can be used. Hence, a physics-based reduction is possible.

In the final step, a sorting algorithm is applied to the reduced basic matrix. Besides the well-known sorting algorithms like RCM and triangular, we, especially, concentrate on weighted sorting algorithms, which account for the weights in a given graph.

The framework returns a permutation vector for the given linear system which can then be used to implicitly reorder the matrix during the linear solving step.

The reordering framework is flexible enough to meet the demands of a **physics-oriented smoothing** within an AMG method not only for the electrochemical simulations considered, but also for other convection-

and/or migration-dominated simulation tasks. We show that the resulting physics-based orderings for the MITReM do much better represent the physics of the underlying PDE system compared to state-of-the-art reordering techniques.

In Section 4.1, we specify terminology from graph theories used in this Chapter. Then, we describe state-of-the-art reordering approaches in the context of AMG and discuss the novelties introduced by the reordering framework in Section 4.2. In Section 4.3, we demonstrate how the basic matrix of the reordering framework is created. We introduce techniques to reduce this basic matrix, in Section 4.4. In Section 4.5, we describe the sorting algorithms used in the framework. Finally, in Sections 4.6 and 4.7 we apply the reordering framework to a convection-diffusion problem and to the MITReM, respectively.

4.1   Terminology

A **finite directed weighted graph** $G = (V, E)$ is a tuple consisting of a finite **vertex** set $V = V(G)$ and an **edge** set $E = E(G)$. Each edge $e_{i,j} = e(v_i, v_j, w_{i,j}) \in E(G)$ has got a starting point $v_i \in V(G)$ and an ending point $v_j \in V(G)$ and is associated with a **weight** $w_{i,j} \in \mathbb{R}$. In the following, we denote i to be the **predecessor** of j and j the **successor** of i. The number of successors of a vertex is called **output degree**, the number of predecessors **input degree**. In the following, we assume all graphs to be finite. Ignoring the weights of a graph leads to an unweighted directed graph, which we call **graph**, in the following.

Each matrix of a linear system can be interpreted as a finite directed weighted graph. To be more specific, each variable i represents a vertex v_i. Each off-diagonal non-zero entry a_{ij} of the matrix represents an edge $e_{i,j}$ of the graph with weight $w_{i,j} = \|a_{ij}\|$. If $a_{ij} < 0$ the edge $e_{i,j}$ is an inflow of vertex v_i and an outflow edge of vertex v_j.

A **subgraph** $S = (V', E')$ of a graph G, is a graph such that $V'(S) \subseteq V(G)$ and $E'(S) \subseteq V'(S) \times V'(S)$. We define a **path** p to be an ordered list of vertices $p = v_1, v_2, ..., v_n$, $n > 1$, with $v_i \in V(G)$ which are connected by edges $e_{i,i+1} \in E(G)$, where $i = 1, ..., n-1$. The **weight** of p is defined by $w_p = \sum_{i=1}^{n-1} w_{i,i+1}$, and its **length** by $l_p = |p| = n$, respectively. If $v_1 = v_n$ the path p is called **cycle**. A graph without any cycle is called **acyclic**.

A **strongly connected component** of a directed graph is a subgraph in which a path from each vertex to every other vertex of the subgraph exists. In particular, this means that it includes cycles.

Assume the set of vertices of a graph $G = (V, E)$ is partitioned into disjoint subsets $\mathbb{P} = V_1, ..., V_n$ such that $V_1 \cup ... \cup V_n = V(G)$ and $V_1 \cap ... \cap V_n = \emptyset$. Then, we can create a graph $Q = (V', E')$ such that each subset $V_i \subset V(G)$ corresponds to exactly one vertex $v_i' \in V'(Q)$, and there exists an edge $e_{i,j}'$ if and only if there exists an edge $e_{k,l} = (v_k, v_l)$ with $v_k \in V_i$ and $v_l \in V_j$. The graph Q is called **quotient graph** of G with respect to $\mathbb{P}$.

4.2   State-of-the-Art Reordering in the Context of AMG

In the mid 90's, many people in the groups around Hackbusch and Wittum where concerned with reordering convection-dominated problems to make AMG more efficient. Corresponding publications focused on scalar systems [5, 54, 86, 107].

A detailed overview is given in Le Borne [62]. Here, basic matrix reduction techniques are introduced as well as reordering techniques for cyclic and acyclic graphs which can also be found in earlier papers, cf. Gutsch and Probst [43], Hackbusch, Gutsch, Maitre, and Musy [46]. In particular, various triangular and feedback vertex set algorithms. Also, ideas on discretization dependent orderings are discussed and the idea of reordering on all levels is introduced. The work investigates backward and symmetric Gauss-Seidel smoothing as well as ILU smoothing for convection-diffusion problems and the Stokes-equation. The Stokes equation is solved by a Frobenius complement approach. The results indicate that in the case of acyclic convection and an suitable ordering of variables no convergence issues occur. Considering cyclic convection this is not the case anymore. However, reordering helps to improve the convergence factor in this case considerably, too.

Kim, Xu, and Zikatanov [57, 58] develop and present an aggregative multigrid method which is convergent for convection-diffusion equations without any constraint on the coarse levels, by applying a triangular sorting algorithm in hand with Gauss-Seidel smoothing. The idea is to first "strengthen the asymmetry" and then applying Tarjan's algorithm to create a triangular ordering. Note that this algorithm is very similar to previous works, especially to [5, 62]. Unfortunately, the authors do not relate their work to former publications in the reordering context. Additionally, we believe that convergence issues may occur for this technique on highly irregular grids.

In Pollul and Reusken [84], reordering techniques are applied to a block Gauss-Seidel (BGS) relaxation in order to solve the compressible Euler equations. The point-coupling matrices (see Section 3.2) build up the blocks for the Gauss-Seidel method. The authors do not use an AMG approach, however, this is the first approach for a point-based solver using reordering techniques. The work suggest to create a smaller dimensional system similar to the "primary matrix" known from PAMG to define a new ordering. In that way, the matrix is reordered point-wise. The authors suggest the use of weights within the triangular sorting algorithm. Results show an improved robustness of the Gauss-Seidel method even for large CFL numbers.

In Section 4.7, we show that the state-of-the-art reordering approaches do not lead to satisfying orderings, especially in areas of the domain where the grid is very unstructured. This is even true if the reduction techniques and/or weights are used. Hence, we suppose a physics-based reduction technique which bases on the information of a vector field. This vector field can either be given externally or computed algebraically. If we base the ordering on such a vector field, which describes the dominant direction of flow, the resulting ordering is very structured even for irregular grids.

4.3   Basic Matrix

The sorting algorithm uses the (reduced) basic matrix for creating the permutation vector. Hence, this matrix should be chosen with respect to the ordering aimed at. We distinguish between the following categories of applications

- scalar applications,

- weakly coupled system applications, and

- strongly coupled system applications,

where strongly coupled means that the couplings between different physical unknowns are strong.

In the scalar case, the basic matrix might be chosen as the original one, however, it is often more useful to base it on a reasonable auxiliary problem with the same number of variables. If considering, for example, the convection-diffusion equation to be solved by a classical AMG method using Gauss-Seidel smoothing, it is beneficial to order the variables with respect to the flow, i.e., according to the convection. Hence, a good choice for the basic matrix would be the matrix which only contains the convective terms. However, in order to define such an auxiliary problem, usually extra information (e.g., coordinates) has to be available.

For weakly coupled applications, it is beneficial to choose the basic matrix separately for each unknown function. To be more specific, we suggest to create a different permutation for each unknown. In this way, the physical properties of each unknown are treated individually by the reordering framework. The basic matrices have the dimension of the number of points where the respective unknown exists and their entries are based on the respective unknown-to-unknown couplings or on an auxiliary matrix similar to the case of scalar applications.

Remark 4.1. *When we want to apply AMG for weakly coupled applications we usually choose unknown-based algebraic multigrid (UAMG), see [29, and references therein] for further details. The UAMG approach creates a different hierarchy for each unknown function. Hence, a different ordering of variables for each unknown function makes sense.*

In the case of strongly coupled system applications, the dimension of the basic matrix should be "number of points" × "number of points". This choice is similar to the concept of the primary matrix in the case of PAMG. With this concept, all unknowns at a point stay together. This is done to preserve the system structure of the matrix, since a reordering strategy which proceeds in a "scalar" fashion would destroy it. This would potentially lead to neglecting the strong inter-unknown couplings.

There are two essential possibilities to create the lower dimensional system for the case of strongly coupled system applications. One way is to choose an auxiliary problem as in the scalar case. The other way is to derive the basic matrix from the original one with the same techniques the primary matrix is derived in the case of PAMG.

Remark 4.2. *Except for their dimension, the primary matrix of a PAMG approach, and the basic matrix of the reordering framework in the case of strongly coupled system applications are usually chosen differently. The primary matrix builds the basis for the coarsening strategy of PAMG, whereas the basic matrix of the reordering framework is used to create permutation to be used within the smoothing.*

We discuss the choice of the basic matrix for scalar applications in Section 4.6. In Section 4.7, we investigate several choices of the basic matrix and their effect on the final permutation for the MITReM.

4.4 Reduction Techniques

The basic matrix can directly be used as a basis for the sorting algorithm to be applied. However, the matrix may contain entries which lead to an unsuited permutation. This is especially the case if the basic matrix is derived from/the same as the original matrix. In this case a reduction of the matrix often helps to improve the permutation.

There are two categories of reduction techniques which can also be used in combination

- algebraic reduction, and

- physics-based reduction.

Algebraic reduction techniques reduce the basic matrix by simple algebraic criteria. We introduce common techniques in Section 4.4.1. In Section 4.4.2, we introduce the new physics-based reduction techniques. We show how physical properties from the original matrix can be used to reconstruct vector fields which describe the flow directions of the ions described by migration and convection. Then, we introduce the physics-based reduction technique which uses such vector fields to construct a reduced basic matrix.

4.4.1 Algebraic Reduction

There are many possibilities to reduce the number of non-zero entries of a basic matrix $B = (b_{ij})_{i,j=1,n}$ with algebraic techniques. These are the most common variants:

1. Strengthening of Asymmetry:
   Drop b_{ij}, $i \neq j$ if $|b_{ij}| \leq c|b_{ji}|$ with $0 < c \leq 1$. In the case of $|b_{ij}| = |b_{ji}|$ and $c = 1$ both couplings are dropped because there exists no dominant direction.

2. Strong Connectivity:
   Drop b_{ij}, $i \neq j$ if $|b_{ij}| \leq c \max_{k,k\neq i} |b_{ik}|$ with $0 < c < 1$.

3. Row and Column Strength:
   Drop b_{ij}, $i \neq j$ if $|b_{ij}| \leq c\max(\max_{k,k\neq i}|b_{ik}|, \max_{k,k\neq j}|b_{kj}|)$ with $0 < c < 1$.

4. Relative to the Diagonal:
   Drop b_{ij}, $i \neq j$ if $|b_{ij}| \leq c|b_{ii}|$ with $0 < c \leq 1$.

Note that c has to be chosen with care and in a way that the relevant information remains in the matrix, so that the sorting algorithm to be applied creates a reasonable permutation. We describe these techniques in detail and give some hints on the choice of c.

Strengthening of the asymmetry (1) is commonly used in the case of convection diffusion problems on a regular grid, where the convection is discretized with upwind differences and the diffusion with central differences, see, e.g., [57]. In this case, the sparsity pattern is symmetric. Applying the method leads to a sparsity pattern which is similar to pure convection. Hence, a flow based ordering can then be easily computed.

In the case of strong connectivity (2), c should be chosen between 0.25 and 0.5, i.e., similar to classical AMG's definition of strong connectivities. If c is chosen larger this would most probably result in the loss of important coupling information. In this case, the resulting matrix would be an unsuited basis for an sorting algorithm.

The row and column strength criterion (3) is a generalization of (2). The row and column strength criterion promises good results also in the case of different flow regimes in the same model.

Using the criterion which drops entries based on their relative size to the diagonal (4) might also lead to inaccurate representations of the flow since strong connections might be dropped, especially, if the matrix considered is not diagonally dominant. In this case, criterion (1) and (3), promise better results.

If we want to combine several of these criteria we will proceed as follows: first, one of the criteria is applied, and all entries which are candidates of being removed are marked. In the next step, the second criterion is applied and all entries to be removed are additionally marked. In the end, all marked entries are removed from the matrix. For example, it might be useful to combine criteria (1) and (4) if the sparsity pattern is not symmetric.

The algebraic reduction techniques introduced are generalizations of the reduction techniques proposed in [62], [84], and [57]. For further comments on the referred sources see Section 4.2.

The basic matrix should represent important couplings for the ordering aimed at. To be more specific, if if we want to compute an ordering of variables in the direction of convection, the basic matrix has to represent the convective couplings of the original matrix. A good representation is, for example, that the sparsity pattern of the basic matrix is the same as the sparsity pattern of pure convection. In the case of a bandwidth reducing ordering like RCM which is usually used with ILU-type smoothers, it should represent all entries in order to reduce the fill-in of such smoothers.

4.4.2   Physics-Based Reduction

The physics-based reduction techniques make use of explicitly given or derived physical information. By explicitly given information, we mean that additional information is provided, for example, in form of an auxiliary matrix, a vector field or a coordinate vector. Note, in many applications, the physical information cannot be derived from the basic matrix without further information.

We distinguish two main categories of physical information, namely, matrix-based and vector field based. The matrix-based reduction techniques make use of additional matrices. The additionally given matrices usually contain a certain physical effect. This information is then used to reduce the basic matrix. A common example is to consider a splitting of the original matrix into its components, like convection, diffusion, reaction, etc. In this case, entries might be removed if, e.g., the convection compared to the diffusion is small. However, since the reduction criteria are strongly dependent on the system considered and very similar to the algebraic reduction techniques, we do not go into further detail. Details can be found in [62].

First, we demonstrate how a vector field can be derived from a matrix. Then, we introduce the vector field based reduction developed to generate a reduced basic matrix from a vector field.

Derivation of a Vector Field

We want to compute a vector field describing the direction of movement. The aim is to extract this information from a given matrix $A = (a_{ij})$ using as little as needed additional information. Two extraction methods are introduced to find the direction of movement

- average coupling direction (AV), and

- strongest negative coupling (SN).

Both methods only use coordinate information to create the vector fields.

The average coupling direction aims at finding a good compromise of all couplings. The vector field contains a vector for each point i which is computed via the formula

$$\mathbf{v}_i = \sum_{j, i \neq j} a_{ij} * \left(\begin{pmatrix} x_j \\ y_j \end{pmatrix} - \begin{pmatrix} x_i \\ y_i \end{pmatrix} \right), \tag{4.1}$$

where $\begin{pmatrix} x_j \\ y_j \end{pmatrix}$ and $\begin{pmatrix} x_i \\ y_i \end{pmatrix}$ are the coordinates of points j and i, respectively.

Strongest negative coupling means that the vector field $\mathbf{V}$ which shall describe the global movement is determined by the strongest negative coupling of each variable. That is, if the strongest negative coupling of variable i is in the entry a_{ij} we will draw a vector $\mathbf{v}_i$ pointing from i in the direction of j with length $|a_{ij}|$.

Note, in some cases the strongest negative coupling is not unique. In this case either no coupling is taken, that is the vector length in this point is zero, the first coupling tested is used, or the vector is derived by an averaging method as in the case of AV. In the latter case, however, only the strongest negative couplings are taken into account. In this thesis, we always use the second possibility, despite of the fact that the result depends on the initial ordering of variables, in this case.

Generation of a Reduced Basic Matrix

We create a new reduced basic matrix $\boldsymbol{B} = (b_{ij})$ from the vector field and the original basic matrix. There are several possibilities to create the new basic matrix. In the application considered, we aim at orderings in the direction of flow, which are typically derived by a triangular sorting algorithm. If the new basic matrix is cycle-free this sorting algorithm computes a permutation which leads to a triangular matrix. Hence, we want to reduce the number of off-diagonal entries as far as possible. To be more specific, we aim at the creation of a new basic matrix which usually has at most one off-diagonal entry per row. Note, that the resulting matrix is not guaranteed to be cycle-free.

We describe the generation of a reduced basic matrix with the help of a vector field. Assume a vector field and a basic matrix to be given. In order to compute the reduced basic matrix, we replace each vector of the

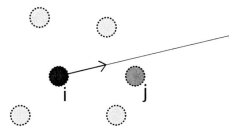

Figure 4.1: Point i, its neighbors and the straight line given via the vector field. Point j with the smallest distance to the line is marked in dark gray.

vector field by a straight line with the same origin and direction. Assume we start with point i. The neighbor point which has the smallest distance to the straight line with origin in i is the only neighbor point which coupling enters in the new basic matrix. Note that the original neighbor points are defined through the couplings of the matrix the original basic matrix. If two point have an equal distance to the straight line the one which is closer to the origin of the vector is chosen. If their distance to the origin is also the same the couplings from i to both points enter the basic matrix. In the case of Figure 4.1, row i has non-zero entries b_{ii} and b_{ij}.

Remark 4.3. *The SN method can also be seen as an algebraic reduction technique. The result can also be achieved by eliminating all couplings in each row of the basic matrix with exception of the strongest negative coupling.*

4.5   Sorting Algorithms

The sorting algorithm is the main part of the reordering framework. The choice of the sorting algorithm depends essentially on two properties. First, the linear solver to be used and second, the type of the basic matrix. When using an ILU-type solver, for example, a triangular ordering of variables is often unsuited, better results can often be achieved using an algorithm which reduces the bandwidth of the matrix.

We distinguish two types of basic matrices, those representing a graph without cycles and those representing cyclic graphs. All algorithms for cyclic graphs can also be used for acyclic graphs However, in this case they are usually equivalent to one of the algorithms for acyclic graphs. It is sometimes also possible to use the algorithms for acyclic graphs in the context of cyclic graphs. However, the resulting ordering most probably depends on the initial ordering in this case. Further details are given in Sections 4.6 and 4.7.

Note that we always do a column and row reordering with the same permutation vector. That is the reordered matrix A' is computed via $A' = P^T A P$, where P is the permutation matrix and A the original matrix. There exist also permutations/matrix manipulations which multiply A with matrices P and Q from the left and right, respectively. P and Q are either diagonal matrices used for rescaling or some kind of permutation matrices. Such algorithms are not considered here since they change A's couplings with potentially negative effects on the convergence of AMG.

We start with a description of the choice of a good starting point/vertex for the sorting algorithms in Section 4.5.1. In Section 4.5.2, we introduce sorting algorithms which are usually used for graphs without cycles. In Section 4.5.3, we describe some extensions to these algorithms and some additional ones which can deal with cycles.

4.5.1   Starting Point

Usually, the new ordering shall be independent of the original one. In most publications cited in Section 4.6, the starting point of the initial ordering and the new ordering are the same. In order to be totally independent of the original ordering, however, this is not suitable.

One possibility to ensure an independent ordering would be the choice of a random starting point, the downside of this is that this point might be not suited for the ordering because it is, e.g., in the middle of the domain. Hence, it is better to choose a starting point on the basis of a measure.

A possibility to find a reasonable starting point is to use the weighted graph of the given basic matrix. Choosing the point with the lowest number of neighbors, and large weights turns out to be a good choice. The low number of neighbors ensures that we take a boundary point, the large weights ensure that the point is important. This idea has already been introduced in [84].

4.5.2   Acyclic Graphs

For acyclic graphs, we consider only the well-known triangularization algorithm. In case of an acyclic graph, this algorithm makes it possible to create an ordering which turns Gauss-Seidel into a direct solver, i.e., all elements of the reordered matrix are in the lower triangle. However, linear systems for most industrial applications are not acyclic, i.e., they have strongly connected components.

In the case of cyclic graphs, however, there are untreated nodes in the end. These nodes are components of cycles. There are many possibilities to handle such cases, we describe some of them in Section 4.5.3.

The **triangularization algorithm** can, for example, be found in [63] it is very similar to the weighted version displayed in Section 4.5.3. Basically, the triangular sorting algorithm starts with a vertex which has no inflow and proceeds with its successors. If one of these successors has no further inflow it is numbered next. Whenever a vertex's predecessors are all numbered, it is numbered next. The same procedure is applied for vertices which have no outflow or which successors have all been numbered; those are numbered with the largest possible number.

If the graph contains cycles, there remain untreated vertices, because all elements in the cycle have a predecessor (successor) which is not numbered. The algorithm is linear in the number of vertices and edges.

4.5.3 Cyclic Graphs

We consider the following algorithms for cyclic graphs

- geometry-based,

- reverse Cuthill-McKee (RCM),

- concentric-cycles,

- block-triangularization,

- heuristic feedback vertex set, and

- weighted-triangularization.

Geometry-based means that the algorithm needs at least the coordinate information. There are many possibilities to choose such an algorithm, for industrial problems the algorithm often has to be tailored to the model and geometry considered.

The RCM algorithm is a bandwidth reduction algorithm. The fill-in of the LU-decomposition is strongly affected by the bandwidth of a matrix. Hence, this algorithm is especially suited for ILU-type solvers and smoothers.

A simple and general approach for cyclic graphs is the use of the concentric-cycles algorithm. This method has been introduced in [45] and is a
straightforward generalization of the triangular sorting algorithm.

The block-triangular algorithm is a further generalization of the triangular algorithm. The resulting matrix has a block-triangular structure, where
each block represents a strongly connected component of the graph. Usually
BGS smoothing is used in this context.

The heuristic feedback vertex set method is also a generalization of the
triangular sorting algorithm. The idea is to cut a minimal set of edges
to make the graph acyclic. Then the acyclic graph can be ordered by the
triangular sorting algorithm.

A further generalization of the triangular sorting algorithm is the introduction of weights which has been proposed in [84]. The idea is to replace
every list of elements in the triangular algorithm by an ordered list, where
edges with larger weights are declared to be more important than those with
relatively small weights. With this technique, the new ordering is nearly
independent of the original one.

Remark 4.4. *Even if the starting point has been chosen with care, the new
ordering might be very similar to the original one. This is especially the
case if the basic matrix has not been reduced (well), because most sorting
algorithms base their ordering solely on the adjacency graph. In order to
avoid the dependency on the original ordering, using the weighted graph
throughout the whole reordering procedure is very beneficial and can easily
be introduced in nearly every reordering procedure.*

There are a lot more sorting algorithms than the ones we consider, e.g.,
skyline-ordering, minimum degree, and level set algorithms which are usually used in the context of direct solvers [34]. Additionally, there exist
many algorithms to reorder the matrices in a way that they can be solved
efficiently in parallel, like coloring algorithms or nested dissection. Such
orderings are not considered here because we only deal with the sequential
case. However, all these algorithms can easily be added to the reordering
framework because of its modular concept.

Geometry-Based Algorithms

The information required to reorder geometrically are at least the coordinates of the grid points. A simple geometry-based ordering is the lexico-
graphic ordering (first x, then y-coordinate) or vice versa. It is also possible

to mark special parts of the domain such as inflow or outflow so that these areas can be identified by the framework. By construction, the resulting permutations are independent of the initial ordering.

Dependent on the information available, it might be possible to create orderings specializing in the discretization chosen, such as surface algorithms [62].

Reverse Cuthill-McKee Algorithm

The efficiency of an ILU(0) method strongly depends on the ordering of the given matrix. The efficiency is strongly related to the number of entries in the matrix $\mathbf{R}$ if we consider a splitting $\mathbf{A} = \mathbf{LU} + \mathbf{R}$ where $\mathbf{A}$ is the matrix to be solved for and $\mathbf{LU}$ its incomplete LU-factorization. In order to minimize the number of $\mathbf{R}$'s entries, it is beneficial to reduce $\mathbf{A}$'s bandwidth by a reordering. Bandwidth reduction is NP-hard. In an appropriate time, we are, hence, only able to find a reasonable compromise and not the perfect ordering.

Minimum degree algorithms can be used to find such bandwidth reducing permutations. A comprehensive overview of these kind of algorithms is given in [39]. The most popular minimum-degree algorithm in the context of ILU-type solvers is the RCM algorithm. The original version without reversing the final ordering has been developed by Cuthill and McKee in 1969 [31]. Modifying the algorithm by reverting the arising ordering was proposed by Alan George. Chan and George [26] show that the RCM algorithm is of linear complexity. Compared with the original algorithm, reverting the algorithm usually shows better results for ILU(0).

The algorithm makes basically use of a breadth first search, where a tree of nodes is created. The parent node is the one with the lowest outward degree. Its child are all neighbors which are not yet in the tree in order of their outward degree, starting with the lowest one. The child nodes are then interpreted as new parent nodes, so that the procedure can be repeated recursively until all nodes are contained in the tree.

Concentric-Cycles Algorithm

For acyclic graphs, the concentric-cycles algorithm equals the triangular sorting algorithm (see Section 4.5.2). However, if untreated vertices remain, i.e., the graph is cyclic, a cycle of the unordered set of vertices is taken off the graph. Then, the triangular sorting algorithm is applied to the remaining unordered vertices. This procedure is repeated until all cycles are cut off the graph. The vertices of the cut cycles maintain their relative ordering and are inserted between the inflow and outflow vertices, see Figure 4.2. Hence, the new ordering depends strongly on the initial ordering.

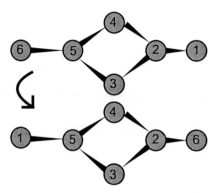

Figure 4.2: Visualization of concentric-cycles algorithm reordering. Initial ordering at the top, new ordering at the bottom.

Block-Triangularization Algorithm

We already pointed out that applying the triangular sorting algorithm directly to cyclic graphs, there are untreated nodes in the end. These nodes are components of cycles, i.e., strongly connected components. In a strongly connected component of a graph, there exists a path from each vertex of the component to every other vertex of the component. Hence, a strongly connected component cannot be ordered in a triangular way.

One possibility to order a graph with strongly connected components is to use the quotient graph, where each strongly connected component (block) is represented by a single vertex. The quotient graph is by definition not strongly connected. Hence it can, e.g., be ordered using the triangularization algorithm. Each strongly connected component of the graph itself is then ordered either in the original ordering or in a new ordering based on weighted graphs. The algorithm is called block-triangularization because it creates a block-triangular ordering, where the blocks are formed by the strongly connected components.

An efficient way to find the strongly connected components of a graph is to use Tarjan's strongly connected components algorithm [101]. Note, in the special case of an acyclic graph, Tarjan's creates a triangular ordering.

Tarjan's algorithm is based on depth-first search (DFS), each strongly connected component in the graph builds up a sub-tree of the DFS-tree.

The algorithm is linear in the number of vertices and edges. During the algorithm, each vertex are assigned two values, a node value and a lowest value of all nodes which can be reached from this node. That is, if these values are the same, the vertex is a root of a strongly connected component.

Heuristic Feedback Vertex Set Algorithm

The feedback vertex set algorithm removes one vertex from each cycle and puts it into a feedback vertex set. The remaining graph is then ordered by the triangular algorithm. The feedback vertex set is numbered at the end. In order to achieve a good ordering the feedback vertex set should be minimized. However, finding a feedback vertex set with a minimal number of edges has proved to be NP-complete [62]. Hence, there exist heuristic feedback vertex set algorithms, see, e.g., [46] (with an idea based on [66]).

Weighed Triangularization Algorithm

The aim of the weighted triangular sorting algorithm is to find a suitable ordering for Gauss-Seidel iteration also in the case of cyclic graphs. The algorithm is proposed in [84].

The weighted version of the triangular sorting algorithm displayed in Figure 4.3 and Figure 4.4 has two lists which are maintained throughout. An inflow ordered list S_i and an outflow ordered list S_o. Outflow/inflow ordered list means that the vertices are sorted using the sum of the weights of their outflow/inflow edges in descending order, starting with vertices that have no inflow/outflow. This results in an ordering where most of the relatively large matrix entries are positioned in the lower triangular part.

4.6   Scalar Example

The ordering of variables may have a strong influence on the convergence rate of iterative (and direct) solvers. The Gauss-Seidel one-level iterative solver, for example, becomes a direct solver if the upper triangular part of the matrix considered is zero. Hence, the ordering of variables of a linear system also effects AMG's smoothing procedure. We consider a simple 2D convection-diffusion example to illustrate the influence of the ordering of variables,

$$- 0.001\Delta u + u_x + u_y = f, \tag{4.2}$$

```
1: Input: vertex set V, number of vertices n
2: Output: Permutation R(old)=new
3: Sᵢ = InflowOrder(V)   !inflow ordered list of V
4: Sₒ = OutflowOrder(V)   !outflow ordered list of V
5: F=1; L=n !initialize counter
6: R=-1 !initialize permutation
7: for k = 1,size(Sₒ) do
8:     p=Sₒ(k)
9:     SetF(p,true)
10: end for
11: for k = 1,size(Sᵢ) do
12:     p=Sᵢ(k)
13:     SetL(p)
14: end for
15: for k = 1,size(Sₒ) do
16:     p=Sₒ(k)
17:     SetF(p,false)
18: end for
```

Figure 4.3: Weighted triangular sorting algorithm - main routine [84].

with Dirichlet boundary condition u=0.

We discretize the PDE with first order upwind finite difference discretization for the convection, second order central finite differences for the diffusion, and a mesh size of $h_y = h_x = \frac{1}{128}$.

The effect of three different orderings to the convergence of the standard AMG method is investigated. We use a random ordering, (x,y)-lexicographic, and (-x,-y)-lexicographic ordering, where (-x,-y) denotes the reversed (x,y)-lexicographic ordering. Furthermore, we set $f = 0$ and use a random first guess so that the initial residual in the L2-norm is 1. The number of iterations to reduce the residual by 8 orders of magnitude with AMG using Gauss-Seidel smoothing, standard interpolation, and a V-cycle without acceleration are shown in Table 4.1. It shows a strong impact of the underlying ordering of variables on the convergence factor.

Note, the different orderings of the variables do not significantly change AMG's coarsening, since it is solely oriented on the strong couplings, which remain in upstream direction. Additionally, the interpolation is pretty accurate in an AMG sense. Hence, the difference between the convergence factors of the different orderings vanishes employing more smoothing steps.

We have demonstrated that the ordering of variables has a significant

subroutine SetF(p,start)
1: **if** $\exists c \in S_o$, c predecessor of p and start **then**
2:     return
3: **end if**
4: $S_i = S_i \backslash \{p\}$
5: $S_o = S_o \backslash \{p\}$
6: $R(p) =$F
7: F=F+1
8: **for** $k = 1$,size(S_o) **do**
9:     v=$S_o(k)$
10:    **if** v successors of p **then**
11:       SetF(v,true)
12:    **end if**
13: **end for**

subroutine SetL(p)
1: **if** $\exists c \in S_i$ c successor of p **then**
2:     return
3: **end if**
4: $S_i = S_i \backslash \{p\}$
5: $S_o = S_o \backslash \{p\}$
6: $R(p)=$L
7: L=L-1
8: **for** $k = 1$,size(S_i) **do**
9:     v=$S_i(k)$
10:    **if** v predecessors of p **then**
11:       SetL(v)
12:    **end if**
13: **end for**

Figure 4.4: Weighted triangular sorting algorithm - subroutines [84].

ordering	iterations	
	1 smoothing step	9 smoothing steps
(x,y)	5	3
(-x,-y)	9	3
random	33	7

Table 4.1: Iteration counts using full Gauss-Seidel smoothing steps.

influence to the convergence behavior of AMG. We describe in which ways the reordering framework can be applied in the case of a convection-diffusion equation to compute a well-suited ordering for Gauss-Seidel smoothing.

In scalar applications, it is possible to use the given matrix as the basic matrix. However, constructing an auxiliary problem often simplifies finding a permutation which suites the linear solvers needs. To further explain the framework for the scalar case, we consider the following linear system derived from a 1D convection-diffusion equation with finite difference discretization and Dirichlet boundary conditions

$$
\begin{pmatrix}
2 & -2 & 0 & 0 & 0 \\
-1 & 3 & -2 & 0 & 0 \\
0 & -1 & 3 & -2 & 0 \\
0 & 0 & -1 & 3 & -2 \\
0 & 0 & 0 & -1 & 3
\end{pmatrix}
\mathbf{u} = \mathbf{f}.
\tag{4.3}
$$

We want to solve this linear system with classical AMG using Gauss-Seidel smoothing. In order to achieve an efficient algorithm, the system has to be reordered such that the convective part is handled well by the smoother, i.e., the lower triangular part of the matrix contains the values -2 and the upper part the entries with -1 as we have demonstrated above.

One option is to define an auxiliary problem. We already pointed out that we aim at an ordering which is beneficial for Gauss-Seidel smoothing. In order to solve the "convective-part" of the error well with help of the smoother, the convective entries should occur only in the lower triangular part of the matrix. Hence, a good auxiliary problem is considering a convection equation. The resulting basic matrix writes

$$
\mathbf{B} =
\begin{pmatrix}
1 & -1 & 0 & 0 & 0 \\
0 & 1 & -1 & 0 & 0 \\
0 & 0 & 1 & -1 & 0 \\
0 & 0 & 0 & 1 & -1 \\
0 & 0 & 0 & 0 & 1
\end{pmatrix}.
\tag{4.4}
$$

Obviously, this basic matrix is cycle-free and can be ordered with the triangular sorting algorithm, which results in the ordering aimed at.

The other option is to define the original matrix to be the basic matrix. The weighted graph of the matrix is visualized in Figure 4.5. If we just consider its adjacency graph (that is each weight is replaced by 1), we are not able to detect the direction of flow. Furthermore, the adjacency graph as well as the original matrix has cycles, for example (1 is connected to

Figure 4.5: Directed weighted graph of the model problem.

2, and 2 is connected to 1). Nearly all of the sorting algorithms proposed return the identity permutation in this case. That is, the new ordering is the same as the old.

There are two strategies to circumvent this problem. One possibility is to reduce the basic matrix before building the adjacency graph, the other one is to make use of a sorting algorithm which uses the weighted graph, and not the adjacency graph, to determine the new ordering. Considering more complex problems, for example with cyclic convection, it might also be beneficial to combine both strategies.

For the model problem, it is sufficient to consider algebraic reduction techniques. The application of physics-based techniques is demonstrated in Section 4.7.

In order to algebraically reduce the basic matrix we could, for example, use the strengthening of the asymmetry criterion with $c = 1$, a reduction with one of the other criteria in this case is also possible (using $c = 0.5$). With these reduction techniques, we strengthen the convective character in the matrix by dropping entries which are regarded as not important in order to find the convection. The reduced basic matrix writes

$$\mathbf{B} = \begin{pmatrix} 2 & -2 & 0 & 0 & 0 \\ 0 & 3 & -2 & 0 & 0 \\ 0 & 0 & 3 & -2 & 0 \\ 0 & 0 & 0 & 3 & -2 \\ 0 & 0 & 0 & 0 & 3 \end{pmatrix}. \tag{4.5}$$

This basic matrix is cycle-free. Hence, the triangular sorting algorithm gives the permutation aimed at.

The alternative to the matrix reduction is the use of a weighted graph within the triangular sorting algorithm, cf. Figure 4.3 and Figure 4.4. In this case, we would create an outflow and an inflow ordered list

$$S_o = 5, 1, 2, 3, 4 \text{ and } S_i = 1, 5, 2, 3, 4. \tag{4.6}$$

The first action takes place in line 17 of the main algorithm, in the called routine we set R(5)=1. Then, the lists are

$$S_o = 4, 1, 2, 3 \text{ and } S_i = 1, 2, 3, 4. \tag{4.7}$$

In the call in line 11 of routine setF nothing happens, since there exists no vertex with zero inflow degree. Hence, the next actions happens again in the call of line 17 in the main routine, where we set R(4)=2. The loop continues and the final result is $R[] = \{5; 4; 3; 2; 1\}$, which is the ordering aimed at.

Summarizing, there are many ways to create the permutation aimed at. The best and most efficient way depends on the application chosen. Sometimes, it is not possible to create an auxiliary problem, or additional information might not be easily accessible. However, even in this case a suitable ordering can often be computed if using algebraic reduction techniques and a weighted algorithm.

4.7   Reordering Framework for the MITReM

The MITReM describes the effects of convection, diffusion, reaction and migration of an electrolyte. Convection as well as migration are effects which describe a "global" movement (flow) of particles caused by external forces. In contrast to this, diffusion and reaction describe local movement which is caused by the concentrations of the ions in the direct surrounding. It is well-known that Gauss-Seidel smoothing shall proceed in the direction of global movement in order to achieve a good AMG convergence. Hence, we evaluate how to create a basic matrix which is well suited to find an ordering taking the effects of migration and convection into account. Additionally, we show how to create a basic matrix suited to achieve a RCM ordering, which is beneficial in the case of ILU-type smoothing.

In the following, we consider the MITReM system, described in Section 2, with parameters stated in Table 4.2.

We discuss the application of the reordering framework for the MITReM as an example of a strongly coupled system of PDEs. In Section 4.7.1, we show how to create basic matrices. Then, we show how to reduce these matrices in Section 4.7.2. Finally, we apply sorting algorithms and show the resulting orderings in Section 4.7.3.

ion	z_i	D_i
$NaS_2O_3^-$	-1	6.00E-10
NO_3^-	-1	1.90E-09
$AgS_2O_3^-$	-1	6.00E-10
Na^+	1	1.33E-09
$S_2O_3^{2-}$	-2	1.90E-09
$Ag(S_2O_3)_2^{3-}$	-3	6.00E-10

Table 4.2: Configuration of the ion-system 3.

4.7.1   Basic Matrix

We use the following basic matrices to compute flow-based orderings

- Frobenius-norm based,

- maximum-norm based,

- auxiliary problem, and

- basic unknown.

The entries b_{ij} of the Frobenius-norm based basic matrix $\mathbf{B}_F$ and the maximum-norm based basic matrix $\mathbf{B}_{max}$ are computed via

$$b_{ij} = \|\mathbf{A}_{(i,j)}\|, \qquad (4.8)$$

where $\mathbf{A}_{(i,j)}$ is the point coupling matrix of points i and j.

We consider several basic matrices which are derived from auxiliary problems. Each of the auxiliary problems represents certain physical effects of the MITReM.

The auxiliary problems are created through a splitting of the original matrix in the components of convection (C), diffusion (D), migration ($M1$ and $M2$), reaction (R), and Poisson equation (P).

$$\mathbf{Ax} = (\mathbf{A}_{M1} + \mathbf{A}_{M2} + \mathbf{A}_D + \mathbf{A}_C + \mathbf{A}_R + \mathbf{A}_P)\mathbf{x} = \mathbf{b}, \qquad (4.9)$$

where $\mathbf{A}_{M1}, \mathbf{A}_{M2}, \mathbf{A}_D, \mathbf{A}_C, \mathbf{A}_R, \mathbf{A}_P$ correspond to parts of the PDE System as follows

$$\frac{\partial c_i}{\partial t} = \underbrace{F z_i u_i \nabla c_i \nabla \Phi}_{M1} + \underbrace{F z_i u_i c_i \Delta \Phi}_{M2} + \underbrace{D_i \Delta c_i}_{D} - \underbrace{\mathbf{v} \nabla c_i}_{C} + \underbrace{R_i}_{R} \qquad (4.10)$$

$$\nabla^2 \Phi = \left. -\frac{F}{\epsilon} \sum_i z_i c_i \right\} P. \qquad (4.11)$$

More precisely, $\mathbf{A}_{M1}, \mathbf{A}_{M2}, \mathbf{A}_D, \mathbf{A}_C, \mathbf{A}_R, \mathbf{A}_P$ is the part of the matrix $\mathbf{A}$ which is described by $M1, M2, D, C, R, P$, respectively.

We consider each of the matrices $\mathbf{A}_{M1}, \mathbf{A}_{M2}, \mathbf{A}_D, \mathbf{A}_C, \mathbf{A}_R$, and $\mathbf{A}_P$ as an auxiliary problem. Additionally, we consider combinations of several of these matrices. Note that the model problems have the same dimension as the original linear system. Since we want a point-wise permutation, they cannot be used directly as a basic matrix. We describe how the important physical information can be extracted from the auxiliary problems and used as a basic matrix.

Assume that the number of ions in the MITReM PDE system is k. Furthermore, assume that the point-coupling matrices are ordered unknown-wise. To be more specific, the row-order of the point-coupling matrices is such that we start with the balance equations for Ions 1 to Ion k and followed by the Poisson equation, and the column-order begins with Ion 1 to Ion k followed by the potential, i.e.,

$$\text{Ion}_1 \quad \cdots \quad \text{Ion}_k \quad \text{Potential} \tag{4.12}$$

$$
\begin{array}{c}
\text{Ion}_1 \\
\vdots \\
\text{Ion}_k \\
\text{Poisson eq.}
\end{array}
\begin{pmatrix}
* & \cdots & * & * \\
\vdots & \ddots & \vdots & \vdots \\
* & \cdots & * & * \\
* & & \cdots & *
\end{pmatrix}. \tag{4.13}
$$

Then, the sparsity patterns of the point-coupling matrices of $\mathbf{A}_{M1}, \mathbf{A}_D, \mathbf{A}_C$ are

$$
\begin{pmatrix}
x & & 0 & 0 \\
 & \ddots & & \vdots \\
0 & & x & 0 \\
0 & & \cdots & 0
\end{pmatrix}. \tag{4.14}
$$

The sparsity pattern of the matrices makes it possible to choose one of the ion-concentrations as a basic unknown to include the relevant information. In the case of $\mathbf{A}_C$, we choose $c(NaS_2O_3^-)$ as basic unknown. Note, since the convection is of the same size for all ions, it does not matter which one we choose in this case. In case of the other matrices $\mathbf{A}_{M1}$ and $\mathbf{A}_D$, however, this is not the case because different basic unknowns have possibly different mobility and diffusion coefficients and a different load. Hence, we investigate several choices of unknowns in these cases.

$\mathbf{A}_{M2}$'s sparsity pattern is

$$
\begin{pmatrix}
0 & 0 & & x \\
 & \ddots & & \vdots \\
 & 0 & 0 & x \\
0 & \cdots & & 0
\end{pmatrix},
\tag{4.15}
$$

and $\mathbf{A}_P$'s

$$
\begin{pmatrix}
0 & 0 & & 0 \\
 & \ddots & & \vdots \\
 & 0 & 0 & 0 \\
x & \cdots & & x
\end{pmatrix}.
\tag{4.16}
$$

$\mathbf{A}_R$'s sparsity pattern depends upon the concrete reactions taking place. Whenever Ion i reacts with Ion j, the point-coupling matrix $\hat{\mathbf{A}}$ has entries in $\hat{a}_{ij}$, $\hat{a}_{ii}$, $\hat{a}_{jj}$, $\hat{a}_{ji}$. Hence, in the fullest case the sparsity patter is

$$
\begin{pmatrix}
x & \cdots & x & 0 \\
\vdots & \ddots & \vdots & \vdots \\
x & \cdots & x & 0 \\
0 & \cdots & & 0
\end{pmatrix}.
\tag{4.17}
$$

Also in the case of $\mathbf{A}_{M2}$, $\mathbf{A}_P$ and $\mathbf{A}_R$, we consider several basic unknowns during the reduction process.

The auxiliary problems are used for a physics-based reduction using flow fields. The aim is to analyze how the framework deals with each of the physical effects in the physics-based reduction procedure. Afterwards, the information is used to create a strategy for the case of a basic matrix which is created via a basic unknown from the original matrix.

4.7.2   Flow-Based Reduction

We show how vector fields can be computed from the auxiliary problems introduced in the previous Section. The vector fields can then be used to create a reduced basic matrices for the sorting algorithms. In the case of the convection auxiliary problem, we use a backward facing step geometry to investigate if its recirculating area is resolved well. For the other problems, we use a channel geometry.

Considerations for $\mathbf{A}_C$

We start with the $\mathbf{A}_C$ matrix, which represents the convection. First, the two methods introduced, strongest negative coupling (SN) and average coupling direction (AV), are analyzed. As stated in the last section, we choose the $NaS_2O_3^-$ ion-concentration as basic unknown.

Figure 4.6 shows the resulting vector fields of the two methods, SN and AV, for a backward facing step geometry, see Section 4.4.2 for details on the methods. The inflow is at the left hand side. Both methods show a very good representation of the convective flow, the main stream as well as the recirculation area at the step are resolved well.

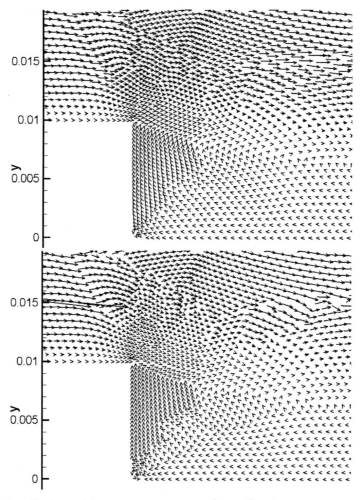

Figure 4.6: Backward facing step geometry (detail). Vector field of $\mathbf{A}_C$ using basic unknown $c(NaS_2O_3^-)$. Computed with AV (top) and SN (bottom).

We already pointed out that the vector fields are the same when choosing a different ion-concentration as basic unknown. The reason for this is that the velocity field used within the convective term of the MITReM is computed by the Navier-Stokes equation and is the same for each concentration unknown. Hence, the choice of a norm-based basic matrix for this auxiliary problem, i.e., replacing each point-coupling matrix by its maximum or Frobenius-norm results in a similar vector field.

Considerations for $\mathbf{A}_{M1}$

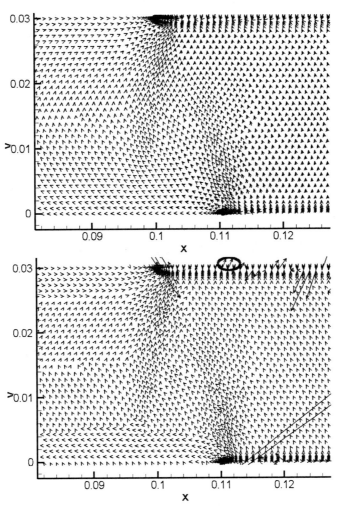

Figure 4.7: Channel geometry (detail). Upper electrode spans from 0.1 to 0.2, lower electrode spans from 0.11 to 1.9. Vector field of $\mathbf{A}_{M1}$ using basic unknown $c(Na^+)$. Computed with AV (top) and SN (bottom).

$\mathbf{A}_{M1}$ represents the part of the migration with a convective character. Figure 4.7 shows the vector fields for $\mathbf{A}_{M1}$ when choosing $c(Na^+)$ as the basic unknown for a channel geometry. Visibly, the averaging method (top) represents physics in a much better way than taking only the strongest negative coupling into account (bottom). Especially at the boundary, SN sometimes shows totally wrong coupling directions and strengths (marked with the circle).

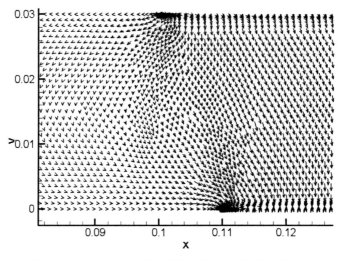

Figure 4.8: Channel geometry (detail). Vector field of $\mathbf{A}_{M1}$ using basic unknown $c(NaS_2O_3^-)$. Computed with AV. Relative vector length.

In contrast to the convection matrix $\mathbf{A}_C$, the choice of the basic unknown in the case of $\mathbf{A}_{M1}$ has a significant effect on the vector fields computed by the two methods. The reason being that the particle velocity which is caused by the migration, depends on the mechanical mobility and the load of each ion. To be more specific, the vector field of a positively loaded ion is in the opposite direction to the one of a negatively loaded ion. This is illustrated by Figures 4.8 and 4.7 (top) which not only show different directions of movement, but also different velocities caused by their different mechanical mobilities.

Considerations for $\mathbf{A}_{M2}, \mathbf{A}_D, \mathbf{A}_R, \mathbf{A}_P$

We choose $c(NaS_2O_3^-)$ as basic unknown to analyze diffusion and reaction. Figure 4.9 exemplary shows the vector fields of the two methods for the diffusion matrix $\mathbf{A}_D$. AV shows only vectors at the boundary. These occur because of Dirichlet boundary points, which were eliminated beforehand. The SN method shows no clear direction because the strongest negative coupling depends on the grid, furthermore it might be not unique. In the latter case, we choose the first coupling of the strongest ones. Of course, diffusion has no global direction of movement, hence, there exists no vector field representing the flow.

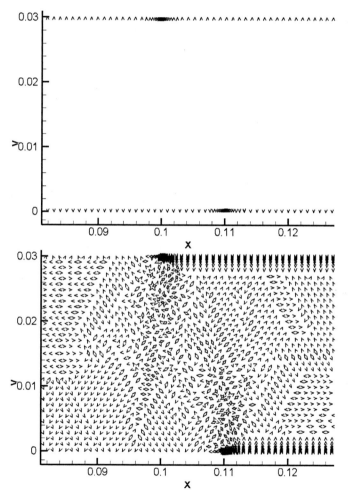

Figure 4.9: Channel geometry (detail). Vector field of $\mathbf{A}_D$ using basic unknown $c(NaS_2O_3^-)$. Computed with AV (top) and SN (bottom).

A possibility to enhance the SN approach for the cases where no global movement takes place is the use of an algebraic reduction method beforehand. If we choose, for example, the strengthening of the asymmetry method ($c = 1$) we end up with a diagonal matrix, because the basic matrix is symmetric. In this case, the result for SN, of course, looks similar to the AV displayed in Figure 4.9.

The vector fields for $\mathbf{A}_R$ are similar. Furthermore, choosing the potential as basic unknown in the case of $\mathbf{A}_P$ leads to similar results as in the case of $\mathbf{A}_D$ with an ion-concentration unknown displayed in Figure 4.9 because this coupling also has a diffusive character.

In the case of $\mathbf{A}_{M2}$, no couplings are on the diagonals of the point-coupling matrices. Hence, choosing an ion-concentration or potential unknown would lead to a zero matrix. It is also possible to choose, for example, the coupling of an ion-concentration to the potential. In the case of $\mathbf{A}_{M2}$, this leads to a diffusion-type matrix and, hence, to similar results as shown in Figure 4.9.

Combinations of Physical Effects

Summing up the matrices $\mathbf{A}_{M2}, \mathbf{A}_D, \mathbf{A}_R, \mathbf{A}_P$ with one of the matrices $\mathbf{A}_{M1}$ or $\mathbf{A}_C$, and choosing an ion-concentration as basic unknown would lead to similar results to the ones presented for $\mathbf{A}_{M1}$ and $\mathbf{A}_C$, respectively when using AV. However, in the case of SN the result strongly depends on the size of the physical effects involved.

Combining all effects, i.e., considering the original matrix and choosing an ion-concentration as basic unknown, the result strongly depends on the strength of the migration and convection part. In the case of the MITReM, the flow velocity is usually chosen relatively large to satisfy an uniform distribution of the ion concentrations. Hence, the convection is the dominant part. Therefore, the vector fields for the original system look similar to the case of $\mathbf{A}_C$ (for AV as well as for SN) except for the part of the domain with strong migration which is near the electrodes.

Choosing the potential as basic unknown leads to a matrix with diffusive character, because neither migration nor convection is present in these couplings. Hence, in this case, the vector fields are the same as in the case of $\mathbf{A}_D$.

We conclude, that the average coupling direction shows the best results. However, it makes use of the coordinate information, which has to be available. The strongest negative coupling only leads to a representative vector field if a flow direction is dominant.

We have shown how to extract the velocity fields out of basic matrices. The vector fields can, then, be used to set up reduced basic matrices for the sorting algorithms.

Remark 4.5. *Considering only the original matrix without additional information, it is usually not possible to derive matrices from the original matrix which describe selected physical effects. However, in the case of the MITReM, we are able to separate the effects of convection and migration. the reason being that the flow caused by the convection for each ion-concentration is the same, and the flow caused by migration depends on the ion properties.*

Let $\mathbf{B}(c_i, c_i) = \mathbf{B}_i + \mathbf{C}_i + \mathbf{D}_i + \mathbf{R}_i$ be the basic matrix created by using basic unknown c_i with migration part $\mathbf{M}_i$, convection part $\mathbf{C}_i$, diffusion part $\mathbf{D}_i$, and reaction part $\mathbf{R}_i$. Assuming there exist two ions (i and j) which differ in the product of load (z_i) and mobility (m_i), the effects can be separated by subtracting the matrices $\mathbf{B}(c_i, c_i)$ and $\mathbf{B}(c_j, c_j)$ that is

$$\hat{\mathbf{B}} = \mathbf{B}(c_i, c_i) - \mathbf{B}(c_j, c_j) \quad = \quad \mathbf{M}_i - \mathbf{M}_j + \hat{\mathbf{D}}_i - \hat{\mathbf{D}}_j + \underbrace{\mathbf{C}_i - \mathbf{C}_j}_{=0} \quad (4.18)$$

$$= \quad (1 - \frac{m_i z_i}{m_j z_j})\mathbf{M}_i + \hat{\mathbf{D}}, \qquad (4.19)$$

where $\hat{\mathbf{D}} = \hat{\mathbf{D}}_i - \hat{\mathbf{D}}_j$, and $\hat{\mathbf{D}}_k = \mathbf{D}_k + \mathbf{R}_k$, $k = i, j$.

Hence, the difference of these matrices, $\hat{\mathbf{B}}$ includes the information of the migration. The matrix can be multiplied by factors computed through the mobilities and loads of other ions in the ion-system considered, so that the migration field for each ion can exactly be reconstructed.

Multiplying $\hat{\mathbf{B}}$ by a factor which is computed through the load and mobility difference of the two ions considered and subtracting it afterwards from $\mathbf{B}(c_i, c_i)$

$$\mathbf{B}(c_1, c_1) - \frac{m_2 z_2}{m_2 z_2 - m_1 z_1}\hat{\mathbf{B}} = \mathbf{C}_1 + \tilde{\mathbf{D}}, \qquad (4.20)$$

leads to a matrix which includes the convection field information.

4.7.3   Permutations

We have demonstrated that it is possible to extract flow fields from the original matrix as well as from auxiliary problems which represent the directions of global movement. Hence, when using basic matrices which have been reduced with help of the flow fields, we can create orderings which do represent the direction of global movement very well. These matrices are suited to create orderings which are beneficial for Gauss-Seidel type smoothers. If considering ILU-type smoothing, however, usually a RCM ordering is better suited than an ordering which solely takes the flow into account. A suited basic matrix for this case can be computed using a norm-based approach. Generally, the choice of the reordering-type has to be done with respect to the smoother employed.

Notation 4.6. *In the following Figures, the ordering of points is visualized by colors. The point with the lowest index is colored dark blue, and the point with the largest index is colored red. These figures only visualize the quality of the ordering, and not the direction. Studies on the suitability of the governed orderings of variables for the smoothers employed can be found in Section 7.3.2.*

Reverse Cuthill-McKee (RCM)

The original ordering given by the simulator employed is RCM. However, the starting point is in the middle of the lower domain boundary which does not lead to optimal results. Hence, we want to place the starting point at the inflow.

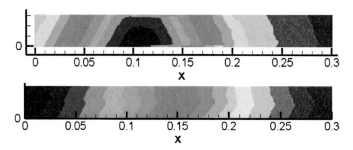

Figure 4.10: Channel geometry. Original RCM ordering (top) and new RCM ordering (bottom). Ordering proceeds from blue to red.

We have already pointed out, that the basic matrix in the case of a RCM type ordering should represent all couplings. Hence, we apply the RCM algorithm to the basic matrix computed with the maximum-norm using no weights, except for the choice of the starting point. In this way the starting point is chosen at the inflow of the domain, see Figure 4.10.

Remark 4.7. *An alternative, would be using an ion-concentration as basic unknown because the sparsity pattern is the same as for the norm-based basic matrix in the case of the MITReM.*

Block-Triangular

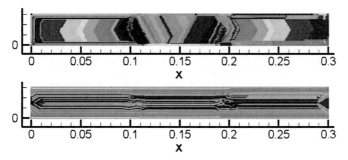

Figure 4.11: Channel geometry. Block-triangular ordering for the migration M_1 (top) and block-triangular ordering for the dominant direction derived with AV method using basic unknown $c(NaS_2O_3^-)$ (bottom). Ordering proceeds from blue to red.

The (block)-triangular ordering is known to improve AMG's robustness and efficiency in case of Gauss-Seidel type smoothing. Figure 4.11 shows results for the block-triangular sorting algorithm. The ordering in the direction of migration is only based on the migrative part of the matrix, which has been gathered as described in Remark 4.5. The basic matrix was reduced via the AV method. The ordering in the dominant direction also uses the AV method and $c(NaS_2O_3^-)$ as basic unknown. That is, it uses the original matrix and additionally coordinate information.

Both methods use weights in order to improve their result. Using no weights gives a very unstructured ordering, see Figure 4.12. Nonetheless, the recirculation area is resolved very well in the case of the backward facing step (BFS) geometry.

Basing the block-triangular sorting algorithm on a basic matrix created by a norm-based approach gives rather unstructured results because several physical effects are mixed-up. For example, taking a Frobenius-norm based basic matrix without reduction and using the block-triangular reordering which employs weights leads to the ordering shown in Figure 4.12. We observe that the result does not represent the flow direction, especially, in the area where the grid is very unstructured.

General Comments

We conclude that, two aspects are important in order to produce a permutation that reasonably well represents a flow. First, the basic matrix has

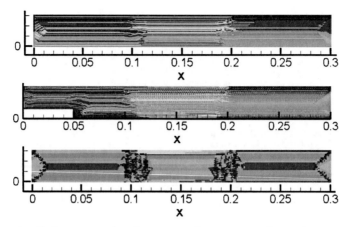

Figure 4.12: Ordering in the dominant direction with neither weights nor reduction technique (channel top, BFS middle); ordering in the dominant direction using weights and Frobenius-norm based basic matrix (channel bottom). Ordering proceeds from blue to red.

to clearly represent this flow and, second, it is most beneficial, instead of reducing the basic matrix, to use a weighted block-triangular reordering so that no flow-information gets lost.

Using a norm-based approach usually makes it difficult to recognize the flow direction, since all effects mix up in the basic matrix. That is, at least the coordinate information should be used to achieve a reasonable permutation.

However, in the case of a RCM ordering, there is hardly any difference between a basic unknown approach and a norm-based approach as long as the basic unknown contains all relevant coupling information, i.e., the sparsity pattern of the basic matrices are the same.

Figure 4.13 shows the sparsity patterns of the permuted matrices. Indeed, the RCM ordering produces a very small bandwidth. The two block-triangular orderings of variables spread the matrix entries relatively uniform all over the matrix.

Note that it is of course also possible to choose a different sorting algorithm, e.g., the heuristic feedback vertex set algorithm (HFVS) proposed in [62]. However, the HFVS algorithm has a higher computational cost and does not lead to significantly better results in the case of the MITReM. Hence, we only use the block-triangularization in case of Gauss-Seidel smoothing. The block-triangularization shows the best overall performance with respect to run time and convergence factor of the AMG

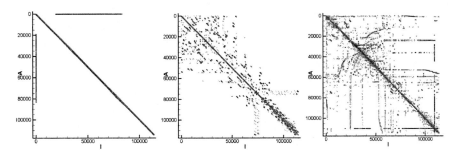

Figure 4.13: Channel geometry. Sparsity pattern of RCM ordering; block-triangular ordering in direction of migration; block-triangular ordering in the direction of convection (left to right).

approach for the chosen basic matrices.

Chapter 5

The Heuristic Péclet Number

The Péclet number describes the ratio between convection, diffusion and mesh size. Hence, it can only be determined when the coordinate and grid information is available and the convection and diffusion in each grid point is known.

In Section 2.6, we have pointed out that a discretization with finite elements using linear shape functions for convection-dominated problems leads to checkerboard instabilities if the Péclet condition is not fulfilled. Furthermore, we have described that the *migration* term, which is, numerically, very similar to the convection, uses this type of discretization. Hence, we checkerboard instabilities arise for the MITReM if the Péclet condition is not fulfilled.

Furthermore, it is well-known that even if the Péclet condition is fulfilled on the first level, it might be violated on coarse levels in the case of "pure" geometric multigrid, cf. Section 3.3. The reason is the construction of the coarse-grid problems. Namely, the fine and the coarse-grid problem are the same except for the different underlying mesh sizes.

Using the Galerkin operator, the linear systems to be solved on the coarse levels are *not* equivalent to the ones of the respective fine level with just a different mesh size. Consequently, we do not know the Péclet number on the coarse levels. Additionally, we have no grid on the coarser levels of the AMG method, instead, only a graph of connectivities is available. Hence, we cannot compute the Péclet number on the coarse levels of the AMG method with the definitions commonly used, cf. Section 2.6.1.

Therefore, we introduce a **heuristic Péclet number** for scalar as well as systems of PDEs which is **purely based on algebraic criteria**. We demonstrate that this heuristic Péclet number is similar to the mesh Péclet number on the fine level. Hence, it allows the **localization of numerically critical part of the domain** caused by checkerboard-type instability on the fine level. In contrast to the mesh Péclet number, the heuristic Péclet number developed does not need any grid information. Thus, it can also be used to locate numerical instabilities on coarse levels.

In the case of the MITReM, the heuristic Péclet number developed takes the convection, migration and diffusion of the PDE system into account. In Chapter 7, we show that the heuristic Péclet condition is locally violated for the electrochemical setups considered. If the Péclet condition is violated, usually, the discretization-type or mesh should be changed. However, the given simulator does not allow a change of the discretization. Choosing a finer mesh is not possible because of memory restrictions. Nonetheless, the solution of the simulation represents the physical solution well enough to be used in industrial processes.

We investigate the effects of a locally violated Péclet condition for the hierarchy of AMG in the case of the scalar convection-diffusion equation. We observe that the heuristic Péclet numbers increase and that the area of violated Péclet condition spreads out on coarse levels when using standard AMG components. In order to prevent this behavior, we suggest a **physics-aware coarsening and interpolation** for the scalar case which takes the information of the heuristic Péclet number into account to localize and remedy the numerical difficulties on all levels of the AMG method. The methods proposed are generalized for the case systems of PDEs and applied to a migration-diffusion PDE system in Chapter 6, and to the MITReM in Chapter 7.

In Section 5.1, we introduce a heuristic Péclet number for the scalar convection-diffusion equation. We consider several methods in order to optimize the choice of C- and F-variables based on the heuristic Péclet number and demonstrate that it is possible to achieve a convergence behavior which is independent from the number of levels employed. In Section 5.2, we generalize the heuristic Péclet number for the case of systems of PDEs, in particular, for the MITReM.

5.1   Heuristic Péclet Number for Scalar Equations

We consider the convection-diffusion equation to investigate the dependency of AMG's convergence factor on the Péclet number. More precisely, we consider

$$- D\Delta u + \mathbf{v}\nabla u = f \text{ in } \Omega, \tag{5.1}$$

with Dirichlet boundary conditions. Furthermore, Ω is a domain in $\mathbb{R}^n$, $D \in \mathbb{R}$, $D > 0$ is the diffusion constant, $\mathbf{v} \in \mathbb{R}^n$ the velocity vector, f a given scalar function, and u the scalar unknown function to be solved for. D and $\mathbf{v}$ are assumed to be constant. The equation is discretized using finite differences on a regular grid with mesh size h.

In Section 2.6.1, we have introduced the mesh Péclet number for this convection-diffusion problem

$$\overline{\text{Pe}} := \frac{h}{D}\|\mathbf{v}\|_\infty. \tag{5.2}$$

However, since industrially relevant applications do not usually deal with regular grids there exist alternatives for irregular grids, for example, the element Péclet number known from finite element approaches, cf. Section 2.6.1.

We are going to compute the Péclet number for all point on each level of an AMG approach in order to localize numerical difficulties. However, on coarser levels, there exists no grid information with exception of the coupling information between the coarse level variables. Hence, we construct an alternative Péclet number which measures numerical difficulties on coarse levels. We call this Péclet number "heuristic Péclet number" and show that it equals the mesh Péclet number on a regular mesh.

Usually, the convection and diffusion coefficient in each point of the grid has to be available for the computation of a Péclet number. However, since we plan to base the heuristic Péclet number only on algebraic information, we only assume that the convective and diffusive part of the matrix are stored separately. The separately stored convective and diffusive parts of the matrix allow us to keep track of the convection and diffusion on coarser levels. To be more specific, let the matrices be $\mathbf{C}$ for the convective and $\mathbf{D}$ for the diffusive part, that is $\mathbf{A} = \mathbf{C} + \mathbf{D}$. Then, $\mathbf{C}^{(l+1)}$ and $\mathbf{D}^{(l+1)}$, i.e., the convection and diffusion on level $l + 1$, can be computed via the Galerkin principle,

$$\mathbf{C}^{(l+1)} = \mathbf{R}^{(l)}\mathbf{C}^{(l)}\mathbf{P}^{(l)} \tag{5.3}$$
$$\mathbf{D}^{(l+1)} = \mathbf{R}^{(l)}\mathbf{D}^{(l)}\mathbf{P}^{(l)}, \tag{5.4}$$

where $\mathbf{P}^{(l)}$ and $\mathbf{R}^{(l)}$ are the interpolation and restriction, respectively, of the AMG approach between levels l and $l+1$.

We define the heuristic Péclet number, $\widehat{\mathrm{Pe}}(i)$, in point i to be

$$\widehat{\mathrm{Pe}}(i) := \begin{cases} \infty, & \text{if } \exists j \text{ such that } c_{ij} \neq 0 \text{ and } d_{ij} = 0 \\ \max_{j \in N_i} \frac{2|c_{ij}|}{|d_{ij}|}, & \text{else} \end{cases}, \qquad (5.5)$$

where $j \in \{1,\dots,n\}$ and n is the number of variables. N_i is the set of all points which are coupled to point i via diffusion (neighbors), i.e., $d_{ij} \neq 0 \Leftrightarrow j \in N_i$, and c_{ij} and d_{ij} are the entries of the convection and diffusion matrix, respectively. This definition *only* requires the splitting into convective and diffusive part for each variable and each level.

Theorem 5.1. *In the case of the convection-diffusion equation considered on a regular mesh, constant coefficients, and central finite difference discretization of second order the heuristic Péclet number equals the mesh Péclet number.*

Proof.

$$\widehat{\mathrm{Pe}}(i) = \max_{j \in N_i} \frac{2|c_{ij}|}{|d_{ij}|} = \frac{2 \max_{j \in N_i} |c_{ij}|}{\frac{|D|}{h^2}} = \frac{2\frac{\|\mathbf{v}\|_\infty}{2h}}{\frac{D}{h^2}} = \frac{\|\mathbf{v}\|_\infty h}{D} = \overline{\mathrm{Pe}} \qquad (5.6)$$

$\square$

In contrast to the discrete Péclet number for the regular finite difference mesh, the heuristic Péclet number is locally defined. That is, it may be different for each grid point. This property is similar to the element Péclet number for finite element discretizations.

Notation 5.2. *The Péclet condition in a point is violated if the heuristic Péclet number in this point is greater than 2. In the following, we call these points "critical" since they usually cause numerical difficulties.*

In Section 5.1.1, we perform numerical experiments considering a fulfilled Péclet condition on the fine level. These are followed by an analysis of the case with violated Péclet condition in Section 5.1.2. Finally, we introduce techniques to deal with problems caused by a locally violated Péclet condition on the fine level in Section 5.1.3.

Notation 5.3. *In the following sections, AMG is used without acceleration. Furthermore, if we speak of one-level AMG methods, we mean the smoother of the AMG method used as an one-level iterative solver.*

In the following tables which show iteration counts, we use a zero right hand side, a first guess vector of ones and a relative residual reduction of 8 digits unless stated otherwise. The abbreviation div. used in the following tables denotes that the respective method diverges.

5.1.1 Fulfilled Péclet Condition

We theoretically analyze the case of a fulfilled Péclet condition and perform some numerical experiments.

Theorem 5.4. *We consider the convection-diffusion equation $-D\Delta u + \mathbf{v}\nabla u = f$ on a regular grid with mesh size h, constant diffusion coefficient $D > 0$, constant velocity $\mathbf{v} = \begin{pmatrix} v_1 \\ \vdots \\ v_n \end{pmatrix} \in \mathbb{R}^n$, $\mathbf{v} \neq \mathbf{0}$, and second order central finite difference discretization. If the mesh Péclet condition is fulfilled the arising linear system $\mathbf{A}$ will be an M-matrix and its symmetric part is positive definite.*

Proof. The diagonal elements are positive. If the Péclet condition is fulfilled we additionally have

- the off-diagonal entries are negative. Hence, $\mathbf{A}$ is a Z-matrix. Without loss of generality $v_i \geq 0$ ($i \in \{1,\ldots,n\}$). An off-diagonal element writes $-\frac{D}{h^2} + \frac{v_i}{2h} = \frac{D}{2h^2}(-2 + \underbrace{v_i h/D}_{<2}) < 0$ or $-\frac{D}{h^2} - \frac{v_i}{2h} = \frac{D}{2h^2}(-2 - v_i h/D) < 0$.

- the linear system is irreducibly diagonally dominant.

Hence, the linear system is an M-matrix [112]. Additionally, we have that the transpose of $\mathbf{A}$ is also diagonally dominant. This gives us that the symmetric part $\mathbf{S} := \frac{1}{2}(\mathbf{A} + \mathbf{A}^T)$ is a symmetric M-Matrix. We conclude that $\mathbf{S}$ is positive definite [112]. □

First, we investigate the Péclet numbers on coarse levels for geometric and algebraic multigrid and, then, show numerical results. The 1D model problem considered uses D=1, C=250, h=1/128 and a zero right hand side. We use VAMG with Gauss-Seidel smoothing using standard coarsening to solve the problem.

The fine level stencil for the problem considered is

$$\begin{bmatrix} -384 & 32768 & -32384 \end{bmatrix}_h. \tag{5.7}$$

The stencil on the second level created by geometric multigrid (GMG) writes

$$\begin{bmatrix} 3904 & 8192 & -12096 \end{bmatrix}_{2h},\tag{5.8}$$

and the respective VAMG Galerkin stencil using the same coarse level variables as in the GMG case is

$$\begin{bmatrix} -384 & 32768 & -32384 \end{bmatrix}_{2h}.\tag{5.9}$$

Obviously, the matrix is no longer an M-matrix on the second level in the case of geometric multigrid. Considering VAMG, the stencil is exactly the same as on the fine level. Hence, it stays the same also on further coarser levels as long as we use the same coarsening strategy, namely, leaving every 2nd variable at the current level.

levels	1	2	3	4	5
no. of variables	129	63	31	15	7
iterations	11	2	2	2	2
$\overline{Pe}$	1.95	3.91	7.81	15.63	31.25
$\widehat{Pe}$	1.95	1.95	1.95	1.95	1.95

Table 5.1: Iteration counts and Péclet numbers for VAMG; required rel. residual reduction used is 1e-16.

Table 5.1 shows VAMG's iterations and the Péclet numbers. $\overline{Pe}$ is the discrete Péclet number introduced in Section 2.6.1. On the coarser levels, $\overline{Pe}$ denotes the Péclet number assuming the fine level problem to be the same as the coarse level problem only with h doubled for each coarser level. Hence, it represents the Péclet number as in the case of using GMG. It is given for comparison to the heuristic Péclet number $\widehat{Pe}$.

The table shows that the convergence of the VAMG method is independent of the number of levels, which we expected, since the stencil and, hence, the heuristic Péclet number $(\widehat{Pe})$ does not change.

Remark 5.5. *Considering convection-dominated flow applications, generally, the Péclet condition can be violated on coarse grids even if the Galerkin operator is used [33]. Hence, in this case often artificial viscosity is introduced on coarse grids, cf. Section 3.3.*

We have seen that VAMG's convergence is independent of the number of levels employed in the case of a fulfilled Péclet condition in 1D, because the stencil is exactly the same on each level. Considering 2D and 3D, usually the coarse level stencil does not stay the same as on the fine level, because the coarsening tends to get irregular. In this case, we can loose the M-matrix property. However, experiments show that the Péclet numbers only increase at the boundaries of the domain. That is, it might be locally violated on coarser levels.

5.1.2   Locally Violated Péclet Condition

If the mesh Péclet condition is not fulfilled, the discrete solution of the problem shows oscillations, cf. Section 2.6.1. Hence, this case is not usually considered. However, in practical applications, a locally violated Péclet condition may occur, for example, when the grid is not locally refined enough. If the Péclet condition is only locally violated the results may still represent the physical solution sufficiently well for industrial applications as in the case of the MITReM considered.

In the case of a locally violated Péclet condition, it is not clear if the VAMG algorithm converges because the linear system is no M-matrix anymore. Furthermore, problems for the smoother may occur in this case because the violated Péclet condition results in a non-diagonally dominant matrix.

The VAMG approach employed uses Gauss-Seidel smoothing, standard interpolation, and a V-cycle. On the coarsest level the direct solver PARDISO is used [53]. We investigate the arising problems caused by a violated Péclet condition in 1D and 2D. In 1D, special attention will be payed to the coarse-grid correction. In 2D, we analyze how the Péclet numbers develop on coarser levels.

1D Case

We consider the convection-diffusion equation (5.1) in 1D with mesh size $h = 1/256$ and diffusion coefficient $D = 1$. The domain Ω is split into two sub-domains Ω_1 and Ω_2. The convection coefficient is

$$C = \begin{cases} 0 & \text{in } \Omega_1 \\ 800 & \text{in } \Omega_2 \end{cases} . \tag{5.10}$$

Ω_2 is positioned in the middle of the domain. The mesh Péclet number in Ω_2 is $\overline{\text{Pe}} = 3.125$.

The respective stencils are

$$\begin{bmatrix} -65536 & 131072 & -65536 \end{bmatrix}_h \qquad \text{in } \Omega_1, \text{ and} \tag{5.11}$$

$$\begin{bmatrix} -167936 & 131072 & 36864 \end{bmatrix}_h \qquad \text{in } \Omega_2, \tag{5.12}$$

when neglecting the boundary points of Ω.

The jumping coefficients of this model problem are non-physical. It is used to imitate the large gradient of the migration near the electrode boundary. With the help of the model problem, we show the effects of a locally violated Péclet condition on the hierarchy of VAMG and its smoother.

Remark 5.6. *In the case of pure diffusion, the VAMG approach employed acts as a direct solver for this problem. Hence, we observe a nearly direct solving behavior for small Ω_2, too.*

We use standard coarsening, and ignore the positive couplings in the matrix. If there are matrix rows with only positive couplings the corresponding point will become a C-point. Then, the stencils on the second level are

$$\begin{bmatrix} -32768 & 65536 & -32768 \end{bmatrix}_h \qquad \text{in } \tilde{\Omega}_1, \text{ and} \qquad (5.13)$$
$$\begin{bmatrix} -167936 & 131072 & 36864 \end{bmatrix}_h \qquad \text{in } \tilde{\Omega}_2, \qquad (5.14)$$

where $\tilde{\Omega}_1$ and $\tilde{\Omega}_2$ are the domains Ω_1 and Ω_2, respectively, excluding their outer points. The stencils for the points at the sub-domain interfaces have different values. Nonetheless, the heuristic Péclet number remains the same as on the fine level for all the coarse level variables in this case.

	levels					
points in Ω_2	1	2	3	4	5	6
0	>500	1	1	1	1	1
1	>500	1	1	1	1	1
3	>500	>500	>500	>500	>500	>500

Table 5.2: Iteration counts for VAMG.

We vary the number of points in Ω_2 to investigate the effect on the VAMG convergence. Table 5.2 shows the iteration counts. In the case of 0 critical points, VAMG is a direct solver. Considering one critical point, this is not true anymore. However, the reduction of the residual in the first iteration is still better than 8 orders of magnitude. When 3 points do not fulfill the Péclet condition the AMG method converges very slowly. A further analysis shows that there exists one negative diagonal entry on level 2 which arises at the sub-domain interface.

In the case considered, the negative diagonal element on the second level can be avoided by forcing the inter-domain boundary-points of each sub-domain to the coarsest level. This ensures that the different stencils of the sub-domains do not interact.

points in Ω_2	levels					
	1	2	3	4	5	6
3	>500	1	1	1	1	1
5	div.	1	1	1	1	1
191	div.	1	1	1	1	1
193	div.	1	1	1	1	div.

Table 5.3: Iteration counts for VAMG with improved coarsening VAMG-FC(b).

In the case of 3 points in Ω_2, forcing variables $127, 131 \in \Omega_1$, and $128, 130 \in \Omega_2$ to exist on all coarser levels gives again a very good reduction of the residual. We call the method which uses forced C-points at the inter-domain boundaries VAMG-FC(b). Table 5.3 shows that VAMG-FC(b)'s convergence behavior is stable when using less than 193 points in Ω_2. For 193 critical points, there exists no pure diffusion stencil anymore on level 5. Hence, the smoother diverges very strongly on this level. Furthermore, the diffusion cannot be corrected sufficiently anymore. This results in divergence of the full-level approach starting from 193 points violating the Péclet condition.

Considering the results of the Gauss-Seidel iteration, we observe divergence starting with 5 critical points. The reason is that 3 points are "free", i.e., have no neighbor point which fulfills the Péclet condition in this case. Additionally, considering Gauss-Seidel's behavior on coarser levels, we observe that plain Gauss-Seidel relaxation applied to the coarse level matrices diverges if there exists a cluster of points which violate the Péclet condition with more than 2 critical points.

Summarizing, we have seen that the Gauss-Seidel solver only converges if the number of critical points within a cluster is smaller or equal than 3 in the case of a Péclet number of 3.125 for these points. VAMG might produce negative diagonal entries on coarser levels, already for the case of one critical point. This can be remedied by using VAMG-FC(b) which forces the inter-domain points to the coarsest level. When there exists no pure diffusion stencil at a coarser level anymore even VAMG-FC(b) is not able to converge.

We investigate the effect of the size of the Péclet number on the convergence and use the VAMG-FC(b) approach in order to exclude the inter-boundary effects.

C in Ω_2	$\overline{\text{Pe}}$	levels					
		1	2	3	4	5	6
700	2.34	>500	1	1	1	1	1
800	2.73	div.	1	1	1	1	1
8000	31.25	div.	1	1	1	1	div.

Table 5.4: Iteration counts for VAMG with improved coarsening VAMG-FC(b). Sub-domain Ω_2 includes 5 critical points.

Table 5.4 shows the iteration counts for various convection coefficients for 5 critical points in Ω_2. The model using 5 critical points shows that Gauss-Seidel method converges despite of the fact that there are more than 3 critical points in the case of a slightly violated Péclet condition. VAMG-FC(b) converges even if the Péclet condition is strongly violated with exception of the 6-level method. The reason is the strong divergence of the Gauss-Seidel smoother on level 5. This behavior is shown in Figure 5.1 which compares the error after 3 Gauss-Seidel iterations on level 4 and level 5.

2D Case

We investigate the effects of a locally violated Péclet condition in 2D. Note that the stencil of the discretized convection-diffusion equation is usually not the same on all levels as it has been in the 1D case. Instead, we expect increasing Péclet numbers in the area of a violated Péclet condition. We analyze if this has a negative effect on the convergence of the VAMG approach.

We consider the convection-diffusion equation (5.1) in 2D on a square domain Ω with regular grid. The mesh size is $h = 1/256$, the diffusion coefficient is $D = 1$ and the convection coefficient is

$$\mathbf{v}^T = (C, C) \text{ with } C = \begin{cases} 0 & \text{in } \Omega_1 \\ 800 & \text{in } \Omega_2 \end{cases}, \tag{5.15}$$

where $\Omega_1 = \Omega \backslash \Omega_2$, and Ω_2 is a rectangular domain with height σ_h and width σ_w positioned in the center of Ω. The mesh Péclet number of the points in Ω_2 is $\overline{\text{Pe}} = 3.125$.

Remark 5.7. *In the case of 2D, the VAMG approach used is no direct solver anymore.*

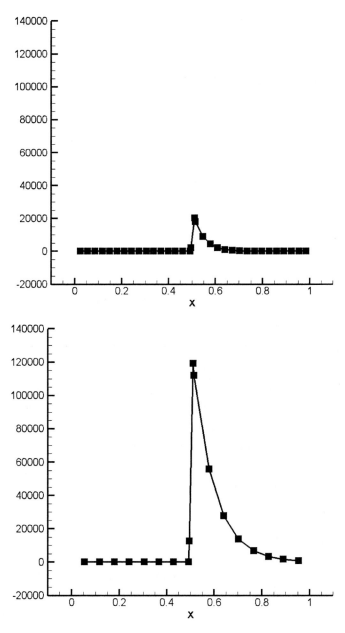

Figure 5.1: Error after 3 Gauss-Seidel smoothing steps on level 4 (top) and level 5 (bottom), starting with a random first guess. Sub-domain Ω_2 includes 5 points with Péclet number 31.25.

Similar to the 1D case, we try to determine the critical size of Ω_2 on a unit square Ω for which the VAMG approach shows convergence problems. First, we use different sized square domains Ω_2. Considering Table 5.5, we see that Gauss-Seidel starts to diverge if there exists more than 9 critical points. That is, there exists more than 1 "free" point which is similar to the 1D case.

	levels				
points in Ω_2	1	2	3	4	5
VAMG					
9	>500	5	7	7	8
25	div.	5	8	8	10
121	div.	10	54	div.	div.
VAMG-FC(b)					
9	>500	5	7	7	8
25	div.	5	6	7	8
121	div.	7	18	10	10

Table 5.5: Iteration counts for VAMG-GS(1,1).

In the 1D case considered, a negative diagonal arises on level 2 of VAMG which leads to a very slow convergence for 3 critical points. This is not the case in 2D. Even for the square including 121 critical points, no negative diagonal arises on coarse levels. Nonetheless, the VAMG approach shows a level-dependency which gets more extreme with increasing number of critical points. For 121 critical points, we do even see divergence.

The level-dependency can be lowered when using a W-cycle instead of a V-cycle. This demonstrates that the smoother is the main reason for the level-dependency. We can remedy the level-dependency by using VAMG-FC(b) which forces the inter-domain points to the coarsest level. Comparing the heuristic Péclet numbers on coarse levels between the methods, we observe that the area including points which violate the Péclet condition spreads out. Furthermore, the Péclet numbers in this area increase. Using VAMG-FC(b), the area stays the same because the interior points of the two domains Ω_1 and Ω_2 do not interact. Additionally, the Péclet numbers do not increase much in Ω_2 and the number of critical points decreases in this case, see Figure 5.2. Hence, the smoother behaves much better on coarse levels in the case of VAMG-FC(b). Note, in the case of VAMG-FC(b), starting from level 4 no interior points in Ω_2 are left.

Comparing the smoother behavior on level 3 between the two approaches,

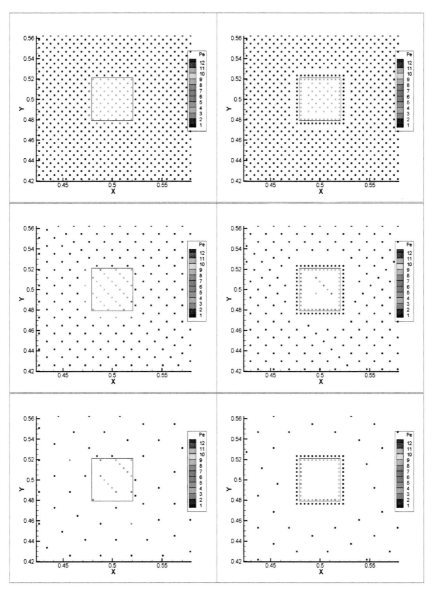

Figure 5.2: Heuristic Péclet numbers on levels 2 to 4 (top to bottom) for
VAMG (left), VAMG-FC(b) (right) for 121 critical points (detail).

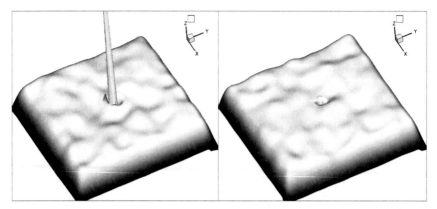

Figure 5.3: Error after 5 Gauss-Seidel steps on level 3 for 121 critical points for VAMG (left) and VAMG-FC(b) (right).

we see that in the case of VAMG a singularity develops, see Figure 5.3. The same can be seen for both methods on level 2. This explains the relatively bad convergence of the three-level method in the case of VAMG-FC(b) and the large iteration count in the case of VAMG compared to the two-level method. This local divergence has been observed in the 1D case, too.

The above results indicate that it might also be possible that Gauss-Seidel method works if the area of violated Péclet condition is relatively wide, but with a low height, and vice versa. To verify this we use different σ_w and σ_h.

σ_h	σ_w	GS	VAMG	VAMG-FC(b)
1	all	> 500	7	7
3	15	> 500	7	7
3	17	div.	7	8

Table 5.6: Iteration counts using various numbers of critical points with height/width $\sigma_{h/w}$ of Ω_2 in points.

Table 5.6 shows that the Gauss-Seidel method is stable if the area of violated Péclet condition is large in width/height and small in height/width. The VAMG approaches behave similar to the case of 9 critical points considering a square domain, cf. Table 5.5. We conclude that the multigrid methods do work sufficiently if the area of violated Péclet condition is large in width/height and small in height/width, too.

Remark 5.8. *If there exist several regions with violated Péclet condition*

*VAMG-FC(b) will behave similar as if there was only one region. The rea-
son is that the variables in the regions violating the Péclet condition do
not interact because of the locally working Gauss-Seidel smoother and the
separation in the coarsening procedure. Considering plain VAMG, it does
not matter how many regions with violated Péclet condition exist as long
as there is enough space in between them so that the areas do not influence
each other on coarse levels and the smoother works locally. However, using
plain VAMG, we see a gradual worsening in VAMG's convergence because
the Péclet number increases on coarser levels.*

We conclude, as long as we have jumping convection coefficients and
the Péclet condition is only locally violated the VAMG-FC(b) method is
an efficient method to solve the arising linear systems. However, it is not
yet clear which strategy shall be chosen in the case of a steady convection
coefficient.

Besides forcing the inter-domain boundary-points to the coarsest level,
there exist also other more general strategies to deal with numerically critical areas. The next section introduces these strategies and investigates their
performance on the model problems considered. Additionally, a new model
problem is introduced which has a steady convection coefficient. We compare the general strategies with VAMG-FC(b) and give some guidelines on
the efficiency of the methods dependent on the model problem properties.

5.1.3   Physics-Aware Coarse-Grid Correction for Problems with Locally Violated Péclet Condition

We have demonstrated that the convergence factor of the straightforward
VAMG method depends on the number of levels in the case of a locally
violated Péclet condition. Hence, we have developed an VAMG strategy,
VAMG-FC(b), which ensures a level-independent convergence factor in the
case of jumping convection coefficients.

There exist several more general strategies to remedy convergence problems which are caused by just a few points. These strategies are often used,
for example, in oil reservoir simulation where the well equations can cause
convergence problems. The strategies introduce a special treatment of the
well equations, so that the convergence of the AMG approach is improved.
We use the heuristic Péclet number to localize the numerical difficulties and,
then, treat the localized variables in a special way. The aim is to develop
an AMG approach which treats areas with violated Péclet condition so that

the convergence factor of the overall AMG approach is independent of the number of levels employed in cases of jumping as well as steady convection coefficients.

We investigate the following common strategies for treating variables which cause convergence problems for AMG

- considering critical variables only within the fine level smoothing process, that is, we do not restrict these variables to the coarse levels (SMO),

- solving a subsystem containing the critical variables with a direct solver and coupling it to the other variables in an alternating Schwarz method fashion (AS), and

- forcing critical variables to exist on all levels (FC).

The first two methods use forced F-point, which means that critical variables are forced to remain on the fine level. Hence, these variables which negatively affect AMG's convergence are decoupled from the hierarchy. It is well known from other applications that decoupling variables which cause problems within the AMG strategy often helps to improve AMG's robustness and convergence [79]. Hence, we apply this kind of decoupling in the case of a locally violated Péclet condition.

For the first method, SMO, it is a prerequisite to have a converging smoother, at least for the critical variables. This strategy very loosely couples the critical variables to the other variables. The method will have benefits if the convergence problems arise only on coarse levels. It is the cheapest of the three methods in terms of computational time, because no extra computations besides the ones to determine the heuristic Péclet number are needed.

In the second case, AS, not only the convergence of the AMG method for the remaining system and the convergence of the alternative method for the critical variables have to be ensured, also the convergence of the Schwarz-type method is important. The convergence of the Schwarz-method depends on the coupling strength between the two sets of variables. We solve the critical variables with a direct solver.

We consider two variants of the AS method, AS(0) and AS(1). The original alternating Schwarz approach AS(0) can be further improved by defining an overlap area so that the neighboring variables of the special variables are treated with the VAMG cycle as well as in the block solve. The overlap used by AS(1) is 1 which means that the neighbors of the critical variables are solved within the direct solve as well as during the VAMG solving step. This approach is slightly more expensive in computational time per iteration compared to AS(0) which uses no overlap region. However, the overlap usually improves the convergence of the method.

In the third case, FC, the aim is to avoid increasing Péclet numbers on coarser levels. The heuristic Péclet numbers are computed on every level. Variables which violate the Péclet condition are forced to exist on the coarsest level. In order to analyze the Péclet numbers on coarse levels two additional Galerkin products have to be computed (one for the convection part and one for the diffusion part of the matrix).

If we force the introduction of additional coarse level variables, this results in shorter distances. Hence, the Péclet numbers increase slower on coarse grids. However, one of two extreme situations might occur. On the one hand, it is possible that the coarsening stops after a few levels, since too many additional C-points are forced. This usually happens if the area where large Péclet numbers occur is large in comparison to the overall domain. On the other hand, additional levels may be introduced because the additional variables on the coarser levels lead to a slower coarsening procedure.

The FC methods force either only the critical variables to exist on all levels (FC(0)), or additionally force the direct neighbors to exist on all coarse levels (FC(1)). FC(0) and FC(1) are very similar to VAMG-FC(b). The difference between the methods is that VAMG-FC(b) does not force the variables in the interior of the area with violated Péclet condition to the coarsest level. Hence, it usually uses less coarse level variables which enhances its efficiency.

Remark 5.9. *In the case of FC and AS, it has to be ensured that the share on the overall number of critical variables is low, otherwise the memory requirements may increase drastically.*

Since, all strategies proposed rely on the information of the heuristic Péclet number, we call the resulting coarse-grid correction physics-aware. We consider two model problems to analyze the performance of the strategies. The first model problem is the 2D convection-diffusion model which we already used to demonstrate the effects of a locally violated Péclet condition on the coarse levels for a straightforward VAMG approach. The second model problem has a steady convection coefficient. The convection direction and strength is chosen similar to the case of the migration in the MITReM.

Notation 5.10. *In the following subsections, the method ORG denotes the* *original VAMG approach.*

In this section, we use the term "convergence factor" which denotes the average reduction of the residual for iterations 3 to N, where N is the number of iterations performed. The first two iterations are skipped, since they usually show an "unnatural" convergence behavior which strongly depends on the first guess. Since we only consider homogeneous problems and stand-alone AMG without acceleration, the governed convergence factor is a good indicator for the asymptotic convergence rate.

Jumping Convection Coefficient

We consider the convection-diffusion equation previously introduced on a 2D square domain Ω with a regular grid with mesh size $h = 1/256$, diffusion coefficient $D = 1$ and convection coefficient

$$\mathbf{v}^T = (C, C) \text{ with } C = \begin{cases} 0 & \text{in } \Omega_1 \\ 800 & \text{in } \Omega_2 \end{cases}, \tag{5.16}$$

where $\Omega_1 = \Omega \backslash \Omega_2$ and Ω_2 is a square domain with height σ positioned in the center of Ω. The mesh Péclet number in Ω_2 is $\overline{\text{Pe}} = 3.125$.

method	levels				
	1	2	3	4	5
ORG	div.	0.46	0.79	div.	div.
SMO	div.	div.	div.	div.	div.
AS(0)	0.97	0.95	0.95	0.95	0.95
AS(1)	0.97	0.91	0.92	0.92	0.92
FC(0)	div.	0.19	0.28	0.43	div.
FC(1)	div.	0.10	0.14	0.18	0.26
FC(b)	div.	0.39	0.47	0.41	0.42

Table 5.7: Convergence rates using a square domain Ω_2 with 121 critical points.

We have pointed out that the heuristic Péclet numbers in the critical areas increase and that the area with critical points spreads out on coarse levels using standard components. This causes convergence problems for the straightforward VAMG approach. We analyze which of the proposed methods SMO, AS, and FC are suited best to remedy these convergence problems.

Furthermore, we have pointed out that SMO is the cheapest one in terms of computations per cycle. It solves for the special variables within the smoothing on the fine level, i.e., no extra computations are needed. However, in order to obtain a converging approach the smoother on the fine level has to converge reasonably fast. We have seen before that the Gauss-Seidel iterative solver for the model problem using 121 points violating the Péclet condition diverges. Hence, SMO diverges, too.

Table 5.7 shows that the AS methods using one level converge. This is achieved because the critical variables are not part of the Gauss-Seidel sweeps, and hence, do not destroy its convergence. Both AS methods show no dependency on the number of levels. However, their convergence for the multilevel approach is only slightly better than for the one-level approach. VAMG for pure diffusion converges at a factor of 0.02, additionally the critical variables are solved by a direct solver. Hence, the reason for the slow convergence is the outer Schwarz method.

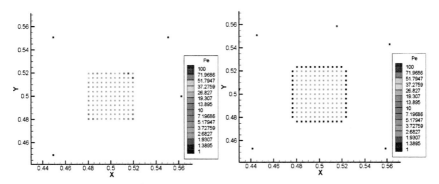

Figure 5.4: Heuristic Péclet numbers on level 5 for FC(0) (left) and FC(1) (right) for square domain Ω_2 with 121 critical points (detail).

The Péclet numbers for FC(0) on coarse levels increase in the critical area, see Figure 5.4 (left). The largest Péclet numbers occur at the boundary and especially at the corners of the critical area. Note that the Péclet numbers of the other variables remain zero. The increasing Péclet numbers can be avoided by additionally forcing the direct neighbors to all coarse levels. In this case, the Péclet numbers on all levels are the same. For FC(0) and FC(1), the Gauss-Seidel smoother diverges on all levels. This is the reason for the level-dependent convergence of these methods. FC(0)'s convergence is even worse than FC(1)'s because of the increasing Péclet numbers on coarse levels, which negatively effect the smoother.

For FC(b), we have observed that the Péclet numbers in the interior of the critical domain increase slowly which is caused by the coarsening process within Ω_2. However, in this case, negative effects are overcompensated by the declining number of critical points, which leads to a converging Gauss-Seidel smoother starting from level 3 for FC(b). Hence, the level-dependency is not observed for this solver. Nonetheless, the convergence factors are worse than for the FC(1) method because not all critical points are solved by the direct solver on the coarsest level.

The results for the methods introduced show that the AS method can enforce convergence. However, if the outer Schwarz method converges slowly, the hierarchy of the VAMG method will not pay-off. The SMO method does not converge at all because of the diverging Gauss-Seidel smoother on the first level. Generally, the FC methods turn out to be the best choice in terms of convergence factor. The factor is nearly independent of the number of levels employed as long as drastically increasing Péclet numbers on coarse levels are avoided as it is the case for FC(1) and FC(b).

The results indicate that FC(1) is the best method in terms of convergence. In contrast to FC(0), it prevent increasing Péclet numbers on coarser levels. This property, leads to a convergent full-level approach and reduces the worsening of the convergence factor with increasing number of levels employed. The drawback of this strategy is a more expensive cycle caused by the additional variables on coarse levels.

In contrast to FC(1), FC(b) forces less points to the coarser levels. Thus, it is slightly cheaper in terms of work per cycle. Furthermore, this method shows also a robust behavior. However, the convergence is not as good as in the case of FC(1), because not all critical variables are solved directly.

Steady Convection Coefficient

The model problem with steady convection coefficient is used to emulate the migration of the MITReM and to find a reasonable strategy which helps to circumvent problems caused by a locally violated Péclet condition within the AMG approach. In the MITReM, the electrodes are usually positioned at opposite boundaries. Their potential difference causes an electromagnetic field between them which drives the migration.

The model problem contains two areas of large Péclet numbers positioned opposite to each other in the middle of the upper and lower boundary of a rectangular geometry. The Péclet numbers increase towards these areas. The convection direction varies and is directed from the lower positioned area to the upper one. Figure 5.5 shows the Péclet numbers and convection direction of this model problem.

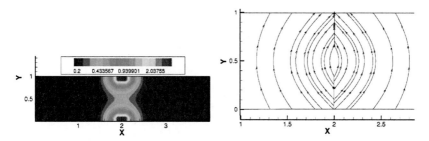

Figure 5.5: Model problem with steady convection coefficient. Péclet numbers (left) and convection direction (right) (detail).

The convection coefficients in x and y-direction, c_x and c_y, are computed as follows:

$$x_{low} = \text{sign}(x-2)\max(\frac{(x-2)^2}{(b/2)^2+b^2},1)$$

$$x_{up} = -x_{low}$$

$$y_{low} = \max((\frac{y}{b})^2,1)$$

$$y_{up} = \max((\frac{(y-1)}{b})^2,1) \qquad (5.17)$$

$$c_x = \frac{2c}{x_{low}+y_{low}} + \frac{2c}{x_{up}+y_{up}}$$

$$c_y = \frac{2c}{|x_{low}|+y_{low}} + \frac{2c}{|x_{up}|+y_{up}}$$

with $c = 800$ and $b = 0.08$. The diffusion coefficient is 1, the length of the geometry is 4 and its height is 1. The mesh size is $1/256$ in both directions.

Similar to the behavior in the case of the model problem with jumping coefficients, the SMO method does not work since the smoother diverges on the fine level. The original method only converges using 2 levels because of increasing Péclet numbers on coarse levels which causes a strong divergence of the smoother.

Table 5.8 shows that the alternating Schwarz-type method converges and that its convergence factor for the two-level method is far better than for the model problem using jumping coefficients, cf. Table 5.7. The reason for the better convergence is the good outer convergence of the alternating method which is caused by the continuous behavior of the convection coefficient. The slight level-dependency of the convergence factor can be explained by the convergence of the VAMG method for the non-critical points which is also displayed in Table 5.8 (last row). The convergence worsens for coarser

method	levels					
	1	2	3	4	5	6
ORG	div.	0.54	div.	div.	div.	div.
AS(0)	0.97	0.02	0.09	0.23	0.21	0.21
FC(0)	div.	0.04	0.03	0.04	0.08	0.07
AS-FC	0.97	0.02	0.06	0.08	0.09	0.10
excluding critical points						
ORG	0.93	0.02	0.06	0.20	0.17	0.17

Table 5.8: Convergence factors for the model problem with steady convection coefficient.

levels since large Péclet numbers arise again. However, since the smoother does not diverge on all levels, in this case, the overall method works well. AS(1) leads to very similar results than AS(0), here, because the outer convergence is very good and cannot be improved much by an overlap. Hence, its results are not displayed in Table 5.8.

The FC(0) method forces all critical variables from the first level to exist on all coarser levels. Within the coarsening new critical points arise, these are also forced to exist on all coarser levels. In the end, the share of forced variables to the number of overall variables is 3.6 percent. The variables on the coarsest level are solved by a direct solver. The remaining points are not critical, and are solved well by the coarse-grid correction. Hence, the FC(0) method shows the best convergence behavior for this model problem and its convergence is nearly independent of the number of levels employed. The use of an overlap does not significantly improve the method.

Table 5.8 also shows the convergence factor for a combination of AS(0) and FC(0) (AS-FC). The method solves the critical variables on the first level within a block solve, only the new critical variables arising on coarser levels are forced to exist on all further coarser levels. This leads to less variables on the coarse levels than in the case of pure FC(0). However, this method only pays in terms of computational time if the additional work on the first level (block solve) is overcompensated by the lower work per cycle on the coarse levels. This is not the case, here, because the convergence is similar to FC(0).

Remark 5.11. *The AS-FC method is equivalent to AS when considering jumping coefficients, since no additional critical points on coarse levels for AS(0) arise.*

Remark 5.12. *We do not show results of FC(b) for the model problem with steady convection coefficient because it is not straightforward to apply this method. In the case of a steady convection coefficient, new variables with large Péclet numbers emerge on coarse levels. One possibility is to force variables only to the next coarser level, and then define new inter-domain points which will be forced next. However, in this case there exist many critical points which will not be solved directly on the coarsest level. Dependent on the size of the critical area, the smoother might diverge strongly which then leads to divergence of the overall method. The other possibility is to act like FC(0) with the difference that on the first level not all critical points are forced to the coarsest level. In this case, the convergence factor is, in most cases, not level-dependent similar to the case of FC(0). However, the convergence is worse than for FC(0), as we have already observed in the case of jumping coefficients. Hence, it is strongly dependent on the model properties if the method pays off in runtime or not.*

Summary

The SMO method diverges because the smoother employed diverges for the critical variables. The AS method works well as long as physical properties of variables within the block solve do not differ much from the other variables (jumping coefficients). However, even in the case where a "continuous behavior" is satisfied the method is only independent of the number of levels employed if the Péclet numbers do not increase on coarse levels. FC shows the most robust behavior, and is nearly independent of the number of levels employed. The only drawback of this method is its higher setup costs and slightly more expensive cycles compared to the original VAMG approach.

5.2 Heuristic Péclet Number for Systems of PDEs

We have demonstrated that a locally violated Péclet condition will lead to divergence of standard AMG in the scalar case. In order to satisfy convergence, the points with violated Péclet condition have to be either decoupled from the hierarchy or the coarse-grid correction has to be changed so that the Péclet numbers on coarser levels do not increase and the area with points violating the Péclet condition does not spread out on coarser levels. The same is expected for the case of systems of PDEs.

We already mentioned that we expect convergence problems of state-of-the-art AMG approaches caused by the discretization of the migration term for the MITReM considered. In order to localize points which may cause checkerboard instabilities, we generalize the heuristic Péclet number for systems of PDEs. Later on, we use the heuristic Péclet number for systems of PDEs to analyze arising numerical difficulties in the case of a migration-diffusion system in Chapter 6 and in the case of the MITReM in Chapter 7.

The residual-distribution N-Scheme which is used for the discretization of the convective term is similar to first order upwind finite differences, see Section 2.6 and references therein. The diffusion and migration terms use a finite element discretization with linear shape functions which is similar to second order central finite differences. Therefore, we first propose a heuristic Péclet number for the scalar case which incorporates a second order central finite difference discretized diffusion and convection term and, additionally, a first order upwind finite difference discretized convection term in Section 5.2.1. Based on this scalar Péclet number, we introduce the heuristic Péclet number for the case of the MITReM and show that a similar definition can also be used for other systems of PDEs in Section 5.2.2.

5.2.1   Scalar Auxiliary Problem

We consider a 1D convection-diffusion equation on a regular grid with mesh size h using a mixed discretization of the convective part. The PDE considered reads

$$- \epsilon u_{xx} + a u_x + b u_x = 0. \tag{5.18}$$

Let $b u_x$ be discretized using first order finite difference discretization, and let the other terms be discretized using a second order central finite difference discretization.

The discrete operator for the first order convection writes

$$\hat{L}_h = \frac{b}{h} [\; -1 \;\; 1 \;\; 0 \;]_h = \frac{b}{2h} [\; -1 \;\; 2 \;\; -1 \;]_h + \frac{b}{2h} [\; -1 \;\; 0 \;\; 1 \;]_h. \tag{5.19}$$

We have transformed the discretization into a second order discretization with an additional viscosity term. Hence, the full operator can be written

$$L_h = \frac{\epsilon}{h^2} [\; -1 \;\; 2 \;\; -1 \;]_h + \frac{a}{2h} [\; -1 \;\; 0 \;\; 1 \;]_h + \hat{L}_h \tag{5.20}$$

$$= \frac{\epsilon + \frac{bh}{2}}{h^2} [\; -1 \;\; 2 \;\; -1 \;]_h + \frac{a + b}{2h} [\; -1 \;\; 0 \;\; 1 \;]_h. \tag{5.21}$$

Consequently, the mesh Péclet number for this problem computes

$$\overline{\mathrm{Pe}}[L_h] = \frac{h(a + b)}{\epsilon + \frac{bh}{2}}. \tag{5.22}$$

Similar to the scalar case considered in Section 5.1, we want to derive a definition of the heuristic Péclet number which only uses entries of the sub-matrices which describe the convection and diffusion. Hence, we consider a splitting of the matrix of the discretized problem to be

$$M = D + C^{(1)} + C^{(2)}, \tag{5.23}$$

where D is the part of the matrix describing the diffusion, $C^{(1)}$ the part of the system describing the convection which is discretized with first order and $C^{(2)}$ the part of the matrix which is discretized with second order.

Then, we can rewrite the mesh Péclet number

$$\overline{\mathrm{Pe}}[L_h] \quad = \frac{2\dfrac{a}{2h} + \dfrac{b}{h}}{\dfrac{\epsilon}{h^2} + \dfrac{b}{2h}} \tag{5.24}$$

$$= \max_{j \in N_i} \frac{2c_{ij}^{(2)} + c_{i,j}^{(1)}}{d_{ij} + \dfrac{1}{2}c_{ij}^{(1)}}. \tag{5.25}$$

We have derived a mesh Péclet number for a 1D convection-diffusion equation which makes use of a mixed discretization of the convective part and which is solely based on the matrix entries of the convection and the diffusion matrix. We discuss the properties of the MITReM and derive a definition of the Péclet number for the MITReM based on the scalar Péclet number for the convection-diffusion equation considered, here.

5.2.2   The Péclet Number for Systems of PDEs

The MITReM describes convection (C), diffusion (D), migration $(M_1$ and $M_2)$, reaction (R) and includes a Poisson equation (P), cf. Chapter 2.

$$\frac{\partial c_i}{\partial t} = \underbrace{F z_i u_i \nabla c_i \nabla \Phi}_{M_1} + \underbrace{F z_i u_i c_i \Delta \Phi}_{M_2} + \underbrace{D_i \Delta c_i}_{D} - \underbrace{\mathbf{v} \nabla c_i}_{C} + \underbrace{R_i}_{R} \tag{5.26}$$

$$\nabla^2 \Phi = -\frac{F}{\epsilon} \sum_i z_i c_i \left.\begin{matrix} \\ \end{matrix}\right\} P. \tag{5.27}$$

After discretizing and linearizing the PDE system, the contribution of the various effects can be located in the linear system. When assuming an unknown-wise order, the coupling structure of the matrix reads

$$
\begin{array}{c}
\begin{array}{cccc}
c(\text{Ion}_1) & \cdots & c(\text{Ion}_k) & \text{Potential}
\end{array} \\
\begin{array}{c}
c(\text{Ion}_1) \\
\vdots \\
c(\text{Ion}_k) \\
\text{Poisson eq.}
\end{array}
\left(
\begin{array}{cccc}
\boxed{M_1{+}C{+}D} & & \boxed{0} & M_2 \\
 & \ddots & & \vdots \\
\boxed{0} & & \boxed{M_1{+}C{+}D} & M_2 \\
 & & P &
\end{array}
\right).
\end{array}
\tag{5.28}
$$

Note that the reaction terms do not appear in the matrix to ensure a good readability. The reaction terms can contribute to the couplings marked with boxes. The exact position of the terms depend on the ions and reactions involved.

Usually, Péclet numbers only contain information about the ratio of convection, diffusion and mesh size. Hence, we neglect the reactive terms in the definition of the Péclet for the MITReM. However, reaction terms may have a drastic impact on the condition of the system. For the MITReM considered, the reaction terms usually lower the impact of a locally violated Péclet condition. We demonstrate this in Section 7.4.

Convective effects are only present in the convection itself (C) and in the convective part of the migration (M_1). Hence, they only contribute to the couplings of a ion concentration to itself, cf. Equation (5.28).

Since the convective effects are only present in the couplings of an ion-concentration to itself, it is sufficient to look at these couplings only. In order to ensure a fast and efficient computation of the Péclet number for the MITReM, the definition of the heuristic Péclet number treats each unknown-coupling separately. That is, the Péclet number of a point is defined to be the maximum Péclet number of its unknown-couplings.

We already mentioned that the terms M_1 and D are discretized using finite elements with linear shape functions, which is similar to second order central finite differences. The remaining effect on the ion-concentration coupling to itself, which effects the Péclet number, is the convection C. The convection is discretized using the N-Scheme of the residual-distribution method which is similar to finite differences of first order.

Hence, we can define the Péclet number in the case of the MITReM analogously to the scalar case. It writes

$$
\widetilde{\mathrm{Pe}}(i) := \begin{cases} \infty, & \begin{aligned} &\text{if } \exists j, l \text{ such that} \\ &d^l(ij) = c^l(ij) = 0 \\ &\text{and } m_1^l(ij) \neq 0 \end{aligned} \\[2em] \displaystyle\max_{j \in N_i} \max_{l \in \mathrm{Ions}} \frac{2m_1^l(ij) + c^l(ij)}{d^l(ij) + \frac{1}{2}c^l(ij)}, & \text{else} \end{cases}
$$

$$(5.29)$$

where N_i is the set of all points which are coupled to point i (neighbors). $m_1^l(ij)$ is the migration part of the $(c(l),c(l))$-coupling of the point-coupling matrix $\boldsymbol{A}_{(i,j)}$ (cf. Section 3.2), and $d^l(ij)$ and $c^l(ij)$ are analogously the diffusive and convective part, respectively. $c(l)$ is the concentration of ion l.

Heuristic Péclet numbers for other systems of PDEs which purely base on matrix-entries can be easily be defined via the scalar case as demonstrated in the case of the MITReM.

Notation 5.13. *Similar to the scalar case, we speak of a **critical point** if the Péclet number at a point is larger than 2.*

Remark 5.14. *In the next chapters, we investigate the Péclet numbers for the case of a migration-diffusion system (Chapter 6) and the MITReM (Chapter 7). In these chapters, we also use similar strategies as in the scalar case to prevent increasing Péclet numbers on coarse levels.*

Chapter 6

A Physics-Aware Point-Based AMG Approach (PAMG) for the Migration-Diffusion System

The MITReM describes many physical phenomena, this makes it difficult to analyze the arising problems properly. Hence, we first simplify the MITReM by only including the migration and diffusion terms. The migration-diffusion system employs the same discretization as the MITReM. Furthermore, it includes the nonlinearities. Hence, the strategies which we develop in this chapter help to develop a PAMG approach for the MITReM.

The migration-diffusion system is also of industrial relevance. Such systems are used for modeling ionic transport processes which are, e.g., very important when simulating MOSFETs (Metal Oxide Semiconductor Field Effect Transistor) which are by far the most common field-effect transistors in both digital and analogue circuits.

Two properties of the migration-diffusion system lead to an insufficient performance of state-of-the-art PAMG methods. These properties are the nonlinearities and the migration dominance near the electrodes.

Due to the **nonlinearity**, the strength of the migration depends on its approximation. Hence, it changes throughout the Newton process. In the model problems considered, the initial gradient of the potential is zero and increases while we approach the solution. On the one hand, this causes increasing Péclet numbers during the Newton iteration. On the other hand, the migration - and hence also the electromagnetic field - for early Newton steps is far from the final one. We show that these properties lead to problems with the PAMG hierarchy for state-of-the-art approaches.

We develop a PAMG approach which uses a **physics-aware primary matrix** to ensure a coarsening which stresses the importance of the migration especially for early Newton steps. Additionally, we generalize the **physics-aware coarsening techniques** to circumvent convergence problems caused by locally violated Péclet conditions. These techniques have already been introduced for the scalar case, cf. Section 5.1.3.

In Section 6.1, we introduce the migration-diffusion system considered. In Section 6.2, we describe the arising problems for state-of-the-art PAMG. The information from the numerical experiments is used to derive a physics-aware primary matrix in Section 6.3. Together with strategies to prevent numerical problems caused by large Péclet numbers introduced in Section 6.4, we, finally, propose an AMG strategy to robustly solve linear systems in Section 6.5.

6.1   The Migration-Diffusion System

The migration-diffusion system considered writes

$$\frac{\partial c_i}{\partial t} = \underbrace{F z_i u_i \nabla c_i \nabla \Phi}_{\nabla_M} + \underbrace{F z_i u_i c_i \Delta \Phi}_{\Delta_M} + \underbrace{D_i \Delta c_i}_{\Delta_D} \qquad (6.1)$$

$$\underbrace{\Delta \Phi}_{\Delta_P} = -\frac{F}{\epsilon} \sum_i z_i c_i. \qquad (6.2)$$

We use the same boundary condition types as in the case of the MITReM, see Section 2.1, and parameters stated in Table 6.1. Note that $u_i = z_i * D_i * F/\Re T$, see Section 2.4. The migration-diffusion system considered is discretized with finite elements using linear shape functions and linearized with a Newton method.

ion	z_i	D_i
Ag^+	1	1.65E-09
NO_3^-	-1	1.90E-09
T		293 Kelvin

Table 6.1: Configuration of the ion-system 1.

The geometry studied is a channel with 36,611 points, see Figure 6.1. Although the channel is a simple geometry, a lot of insight can be gained by investigating the solver behavior for this case. The migration contributions are varying slightly in the geometry and are easy to manipulate by adapting the electrode potential.

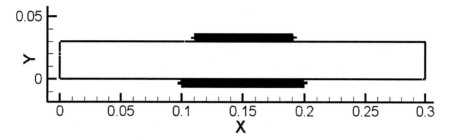

Figure 6.1: Channel geometry with electrodes (gray).

The channel geometry considered has two electrodes positioned opposite to each other in the middle of the channel. The channel length is 0.3, its width is 0.03. The upper electrode's length is 0.1, and the lower electrode has a length of 0.08.

We consider two scenarios which only vary in the difference of the potential between the electrodes. The applied currents at the electrodes for Scenario A are -0.2 and 1.4, for Scenario B the values are -0.2 and 2.1 at the lower and upper electrode, respectively.

Notation 6.1. *In the following, c(X) denotes the concentration of X, where X is one of the ions included in the ion-system considered.*

The coupling structure of the system matrix is

$$
\begin{array}{cc}
 & \begin{array}{cccc} c(\text{Ion}_1) & \cdots & c(\text{Ion}_k) & \text{Potential} \end{array} \\
\begin{array}{c} c(\text{Ion}_1) \\ \vdots \\ c(\text{Ion}_k) \\ \text{Poisson eq.} \end{array} &
\left(\begin{array}{cccc}
\nabla_M + \Delta_D & & 0 & \Delta_M \\
 & \ddots & & \vdots \\
0 & & \nabla_M + \Delta_D & \Delta_M \\
* & \cdots & * & \Delta_P
\end{array} \right).
\end{array}
\tag{6.3}
$$

The matrix has been ordered unknown-wise. Hence, the coupling of an unknown to itself is on the diagonal. We see that the diffusive part of the migration, Δ_M, contributes to the ion-concentration-to-potential coupling. Its convective part, ∇_M, is on the diagonal alike as the diffusion term, Δ_D. Hence, we have a convection-diffusion character for the couplings of an ion-concentration to itself. The heuristic Péclet number which we defined for the MITReM can also be applied for the migration-diffusion system because of the similar coupling structure, cf. Section 5.2.

In the case of the migration diffusion system it writes

$$
\widetilde{\mathrm{Pe}}(i) := \begin{cases} \infty, & \begin{array}{l} \text{if } \exists j, l \text{ such that} \\ d^l(ij) = 0 \\ \text{and } m_1^l(ij) \neq 0 \end{array} \\[2em] \max_{j \in N_i} \max_{l \in \mathrm{Ions}} \dfrac{2m_1^l(ij)}{d^l(ij)}, & \text{else} \end{cases} \qquad (6.4)
$$

where N_i is the set of all points which are coupled to point i (neighbors). $m_1^l(ij)$ is the migration part of the $(c(l),c(l))$-coupling of the point-coupling matrix $\boldsymbol{A}_{(i,j)}$ (cf. Section 3.2), and $d^l(ij)$ is analogously the diffusive part. $c(l)$ is the concentration of ion l.

The Newton method employed uses a static damping. Hence, it reacts very sensitive to the final residuals of the linear solver. If the linear solver is not able to reduce the residual to a certain threshold, the Newton method is likely to diverge. The initial damping factor is set to zero, after each nonlinear iteration the factor is increased by 10 percent until the upper bound of 0.8 is reached.

Scenario A needs 37 non linear iterations, and Scenario B 53 non linear iterations in order to fulfill the residual criterion aimed at. This is due to the fact that the initial guess of the Newton method is chosen very bad, i.e., the start residual in the first Newton step is 10^4 for Scenario A and 10^{10} for Scenario B. Furthermore, the initial guess of the potential gradient is zero, and increases until the final solution of the system is reached. Hence, the Péclet numbers increase for each linear system of the Newton method.

In the following experiments, we use single matrices only to analyze the performance of PAMG. We choose the 5th linear system as a representative for early Newton steps, where the approximation is far away from the solution and the migration is rather low. The matrices are denoted with A-5 and B-5 in the case of Scenario A and B, respectively. Additionally, we use the 30th matrix in the case of Scenario A (A-30), and the 46th matrix in Scenario B (B-46) as representative for late Newton steps where the migration is near the one of the final system.

6.2   State-of-the-Art PAMG Approaches

We use several state-of-the-art PAMG approaches for the numerical experiments in this section. The PAMG approaches differ in the choice of the primary matrix. We investigate the effects of the different coarsening procedures caused by the choice of the primary matrices, especially, we analyze their effect on the size of the Péclet numbers on coarse levels. The findings are used to develop a physics-aware primary matrix in Section 6.3.

In Section 6.2.1 we introduce the PAMG approaches considered. We analyze their convergence results in Section 6.2.2 and observe a strong level-dependency. The reason for this is the locally violated Péclet condition as we will see in Sections 6.2.3 and 6.2.4. Finally, we summarize our findings and give some conclusions in Section 6.2.5.

Notation 6.2. *If we speak of one-level AMG methods, we mean the smoother of the AMG method used as an one-level iterative solver.*

The abbreviation "div." used in the following tables denotes that the respective method diverges.

In this section, we use the term "convergence factor" which denotes the average reduction of the residual for iterations 3 to N, where N is the number of iterations performed. The first two iterations are skipped, since they usually show an "unnatural" convergence behavior which strongly depend on the first guess. Since we only consider homogeneous problems and stand-alone AMG without acceleration, the governed convergence factor is a good indicator for the asymptotic convergence rate.

6.2.1   Approaches

PAMG is described in detail in Section 3.2. The approaches considered make use of BGS smoothing, and single-unknown interpolation (SU), based on the same primary matrix as used for coarsening. On the coarsest level, the direct solver PARDISO is employed. We do not use any accelerator in this chapter in order to clearly see the numerical effects of a violated Péclet condition.

The approach PAMG(max) makes use of a primary matrix based on the maximum-norm, that is, the maximal entry of each point-coupling matrix defines the respective entry of the primary matrix. PAMG(Ag^+) uses the couplings of unknown $c(Ag^+)$ to itself as its primary matrix, and PAMG(NO_3^-) accordingly the couplings of unknown $c(NO_3^-)$ to itself.

Remark 6.3. *The diffusion is the dominant term for all arising linear systems nearly all over the domain. Hence, there is only a negligible effect of the ordering of points on the convergence using the BGS smoother. Therefore, we do not analyze the effect of reordering in this chapter.*

The primary matrices represent the physical effects of the system differently. Investigating the couplings of the point-coupling matrices, we see that in more than 90 percent of the matrices the $c(NO_3^-)$ to potential coupling is dominating. This means that the maximum-norm based primary matrix mainly represents diffusion. The primary matrices based on an ion-concentration primary unknown have a convection-diffusion character.

Figure 6.2 shows detailed information on the entries of the maximum-norm based primary matrix. Each point has a value between 0 and 1, which represents properties of its related row of the primary matrix for systems A-5 and A-30. To be more specific, we count how many times each coupling enters the primary matrix per row and divide it by the number of entries in that row. Hence, a value of 1 means that only this coupling is present in the entire primary matrix's row, and a value of 0 means that the coupling does not enter in the primary matrix in that row. For example, (b) shows the coupling from the Ag^+ concentration to the potential for system A-5, we see that this coupling only enters the primary matrix near the electrode. That is, its coupling is only represented near the electrode in the maximum-norm based primary matrix.

We observe that the $(c(NO_3^-),c(NO_3^-))$-coupling only contributes to the primary matrix at the lower electrode (light blue). The $(c(Ag^+),c(Ag^+))$ coupling contributes to the primary matrix at both electrodes, but only in the case of system A-30, see Figure 6.2 (light blue at the top and orange at the bottom) (e). Hence, it is not present in the maximum-norm based primary matrix for system A-5. We observe a similar behavior for the 5th and 46th linear system of Scenario B. The reason for the dominant behavior of the diffusion-like couplings (c(Ion) to potential) at the electrodes, especially the upper one, for the early systems is caused by the approximation of the potential gradient which is far away from the final solution in this case. This leads to a underrepresentation of the migration in the maximum-norm based primary matrix for early Newton steps.

The primary matrices of all approaches include the flow information which is caused by the electromagnetic field, except for the first linear system where the convective part of the migration is zero because of the initial zero potential gradient. The vector field describing the migration can be

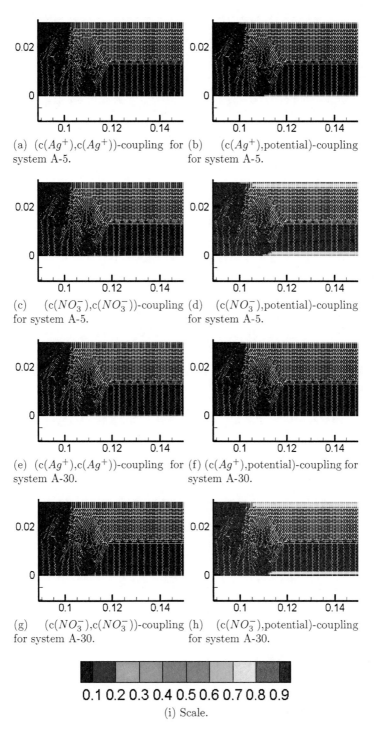

(a) $(c(Ag^+),c(Ag^+))$-coupling for system A-5.

(b) $(c(Ag^+),\text{potential})$-coupling for system A-5.

(c) $(c(NO_3^-),c(NO_3^-))$-coupling for system A-5.

(d) $(c(NO_3^-),\text{potential})$-coupling for system A-5.

(e) $(c(Ag^+),c(Ag^+))$-coupling for system A-30.

(f) $(c(Ag^+),\text{potential})$-coupling for system A-30.

(g) $(c(NO_3^-),c(NO_3^-))$-coupling for system A-30.

(h) $(c(NO_3^-),\text{potential})$-coupling for system A-30.

0.1 0.2 0.3 0.4 0.5 0.6 0.7 0.8 0.9

(i) Scale.

Figure 6.2: Dominance of couplings in maximum-norm based primary matrix at each point of the domain (detail).

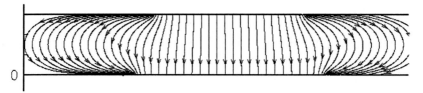

Figure 6.3: Electromagnetic field, directed from top to bottom.

extracted using the AV method proposed in Section 4.4.2. The electromagnetic field is shown in Figure 6.3. It is aligned with the migrative flow of Ag^+ and oppositely directed to NO_3^-. The vector field which can be computed based on the primary matrix of PAMG(max) is similar to the one of NO_3^-.

Summarizing, the PAMG approaches considered differ in the choice of their primary matrices. In every case, the primary matrix represents the diffusion and migration. However, the relations between the effects differ. The approaches based on the ion-concentration primary unknown represent the migration in every point as given. The maximum-norm based primary matrix considers the migration only near the electrodes and nearly neglects it in the remainder of the domain. Additionally, we observe that the migration is underrepresented near the electrodes for early Newton steps when using the maximum-norm based primary matrix because of the bad initial guess.

6.2.2　Convergence Results

We investigate the performance of the PAMG approaches in terms of convergence factors, a special focus is on the level-dependency of the factor. In the scalar case, we observed a significant level-dependency of the convergence factor in the case of a locally violated Péclet condition using a straightforward AMG approach, see Section 5.1.

Table 6.2 shows the convergence factors for Scenario A. PAMG(max)'s convergence factors are stable between 3 and 6 levels for both cases, except for the 6-level method for system A-30, where the approach diverges. In Section 6.2.3, we demonstrate that the divergence is caused by large Péclet numbers which lead to an insufficient coarse-grid correction similar to the investigated scalar case.

Considering PAMG(Ag^+), the convergence factor for system A-5 is not level-dependent. System A-30 shows a slight level-dependency. However,

approach	linear	levels					
	system	1	2	3	4	5	6
PAMG(max)	5	0.90	0.77	0.88	0.88	0.88	0.88
	30	0.96	0.71	0.74	0.74	0.74	div.
PAMG(Ag^+)	5	0.90	0.85	0.88	0.88	0.88	0.88
	30	0.96	0.80	0.83	0.85	0.86	0.86
PAMG(NO_3^-)	5	0.90	0.90	0.89	0.89	0.89	0.89
	30	0.96	0.95	0.95	div.	div.	div.

Table 6.2: Convergence factors for Scenario A.

the convergence factors are worse than in the case of PAMG(max) except for the 6-level approach for system A-30. PAMG(NO_3^-)'s results for system A-5 are in line with the other approaches for system A-5. However, it shows a very bad convergence behavior for system A-30. The divergence of the 4-level method is caused by an inaccurate solution for the coarsest level equations by the direct solver. Additionally, we observe that the BGS smoother starts to diverge beginning on the fourth level.

approach	linear	levels					
	system	1	2	3	4	5	6
PAMG(max)	5	0.78	0.77	0.77	0.77	0.77	0.77
	46	0.88	0.63	0.65	0.65	0.65	div.
PAMG(Ag^+)	5	0.78	0.74	0.76	0.77	0.77	0.77
	46	0.88	0.71	0.74	0.76	0.77	0.78
PAMG(NO_3^-)	5	0.78	0.77	0.77	0.77	0.77	0.77
	46	0.88	0.84	0.83	div.	div.	div.

Table 6.3: Convergence factors for Scenario B.

Table 6.3 shows the convergence factors for Scenario B. The overall behavior of the PAMG approaches is similar to Scenario A. Namely, we observe that for the system from the beginning of the Newton method (B-5), the convergence factor for the approaches is constant. In contrast to this, we observe a level-dependent convergence if we apply PAMG to matrices from the end of the Newton method. This behavior is cause by rising Péclet numbers. We analyze the effect of a locally violated Péclet condition of points on the fine level on the Péclet numbers of points on the coarser levels.

6.2.3   Péclet Numbers on Coarse Levels

We already pointed out that the initial guess for the potential gradient is zero and that its steepness increases during the Newton method. The final potential causes, locally, very large Péclet numbers which lead to problems within the PAMG coarse-grid correction. Nonetheless, the behavior of the approaches differs because of the different coarsening strategies. We investigate the effect of the different coarsening strategies on the size of the Péclet numbers on all levels.

approach	linear	levels				
	system	1	2	3	4	5
PAMG(max)	A-5	14%	20%	27%	36%	14%
PAMG(Ag^+)	A-5	14%	21%	29%	30%	13%
PAMG(NO_3^-)	A-5	14%	21%	30%	35%	28%
PAMG(max)	A-30	15%	20%	27%	36%	25%
PAMG(Ag^+)	A-30	15%	22%	32%	33%	38%
PAMG(NO_3^-)	A-30	15%	22%	32%	35%	38%
PAMG(max)	B-46	17%	25%	36%	50%	30%
PAMG(Ag^+)	B-46	17%	26%	37%	52%	42%
PAMG(NO_3^-)	B-46	17%	26%	37%	53%	45%

Table 6.4: Share of critical points on each level for the various approaches.

Table 6.4 shows the share of critical points, i.e. points which violate the Péclet condition, on each level of the PAMG methods considered. The shares in the case of system B-5 are the same as in the case of system A-5. Hence, they are not displayed. The similarity is caused by the fact that the Newton iteration uses the same initial guess for both systems. Hence, the systems do not differ much at this early point of the simulation.

Generally, comparing the shares in the case of the systems B-46 and A-30, we observe more points with large Péclet numbers on the first level for system B-46, caused by the larger potential difference between the electrodes. Furthermore, the number of points which violate the Péclet condition increases up to more than 50 percent on level 4.

Comparing the PAMG approaches, the shares are very similar (less than 10 percentage points difference) except for level 5. We observe that the approaches using a primary unknown have a larger share of critical points than PAMG(max). However, PAMG(max) diverges using 6 levels where PAMG(Ag^+) converges. Additionally, the shares of critical points on coarse levels for PAMG(Ag^+) and PAMG(NO_3^-) do not differ much, despite of the fact that PAMG(Ag^+) converges and PAMG(NO_3^-) diverges starting from level 4. Hence, we conclude that the share of critical points alone can neither be used to determine convergence factors nor difficulties. This is similar to the findings for the scalar case, where the distribution of the Péclet numbers is more important than their size or share of the total number of points.

In order to find the reasons for the convergence behavior of the PAMG approaches employed, we perform a more detailed analysis of the Péclet numbers and their distribution on coarse levels.

Considering the Péclet numbers of PAMG(max) for Scenario A shown in Figure 6.4, we see a significant difference of the size of the Péclet numbers at the lower electrode between the systems A-5 and A-30 starting from level 4. Namely, the Péclet numbers for system A-5 begin to decrease, so that nearly all points fulfill the Péclet condition on level 5. This is consistent with the observations shown in Table 6.4. Note that the table shows no difference for level 4 because the Péclet numbers at the lower electrode do still violate the Péclet condition for system A-5.

On level 5, we see that nearly all large Péclet numbers disappear for system A-5. In the case of system A-30, critical points also appear in the middle of the domain. Hence, we conclude that the worsening of the convergence behavior between levels 5 and 6 for the linear systems from late Newton steps is caused by the distribution of large Péclet numbers on this level. A further analysis shows that the large Péclet numbers lead to oscillations and divergence for the smoother on level 5 which leads to divergence of the 6-level method for system A-30.

In the experiments for the scalar case, we have seen that the distribution of variables with large Péclet numbers rather than their amount on the overall variables determines the convergence factor of the AMG approach. Hence, we assume that PAMG(Ag^+)'s coarsening anticipates an unsuited distribution of points with large Péclet numbers on coarser levels for the BGS smoother.

Figure 6.5 compares the 5th coarse levels of the three PAMG approaches considered for system A-30. Obviously, at the electrodes the coarsening differs. The PAMG methods based on a primary unknown have more points at the lower electrode than PAMG(max). Additionally, these methods have a larger share of Péclet numbers violating the Péclet condition on this level, which is confirmed by the convergence results shown in Table 6.4. Furthermore, we observe that the distribution of large Péclet numbers is wider for PAMG(max) and PAMG(NO_3^-). Hence, the observation from the scalar case that the distribution and not the share of critical points determines

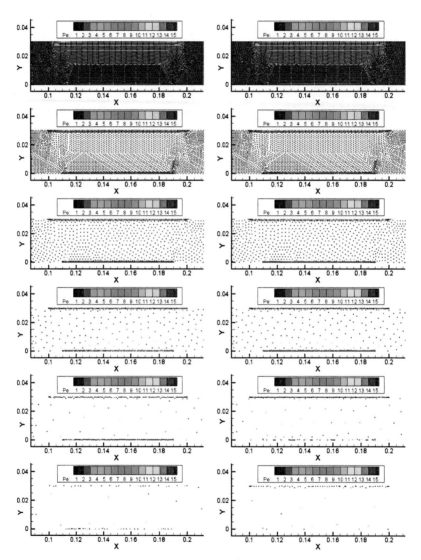

Figure 6.4: Péclet numbers for level 1 to 6 (top to bottom) for linear systems A-5 (left) and A-30 (right) using PAMG(max) (detail).

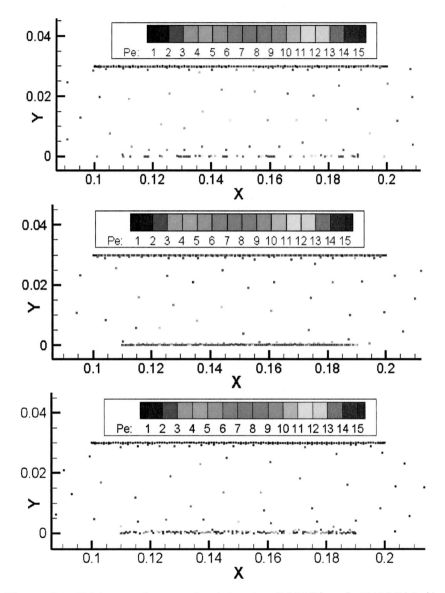

Figure 6.5: Péclet numbers on level 5 using PAMG(max), PAMG(Ag^+), and PAMG(NO_3^-) (top to bottom) for system A-30 (detail).

the convergence of the AMG approach is also observed for system case.

The Péclet numbers in the case of PAMG(NO_3^-) at the lower electrode is lower than in the case of PAMG(Ag^+). However, the distribution of critical points is spread all over the area between the two electrodes on level 5 which is the reason for the divergence of this method.

The Péclet number distribution for Scenario B is similar. However, the Péclet numbers are larger than in Scenario A because of the larger potential difference. Scenario B, shows a similar distribution of large Péclet numbers for the approaches. We conclude that the divergence in the case of PAMG(max) can be explained by the distribution of large Péclet numbers on level 5 which causes oscillations for the BGS smoother for system A-30. The coarsening for PAMG(Ag^+) proceeds in a different way which seems to be better suited for this system. However, comparing the convergence factors of the 5-level methods shown in Table 6.2, PAMG(max) performs better. This effect is not yet explained by the above analysis. The assumption is that the coarsening of PAMG(max) for the part of the domain without electrodes is better suited to achieve a good coarse-grid correction. Hence, we cut the points near the electrodes off the system to verify the assumption.

6.2.4   Ignoring Critical Points

approach	levels				
	1	2	3	4	5
PAMG(max)	0.96	0.55	0.60	0.61	0.61
PAMG(Ag^+)	0.96	0.52	0.61	0.66	0.68
PAMG(NO_3^-)	0.96	0.97	0.97	0.97	0.97

Table 6.5: Convergence factors for system A-30 excluding points near the electrodes.

We extract the subsystem of system A-30 which contains all points with exception of the points near the electrodes. To be more specific, all points within a rectangle of 0.005 around the electrode boundaries are dropped. This means, especially the points which have very large Péclet numbers, and their direct surrounding are not considered. Hence, the resulting fine level matrix has no critical points anymore.

The convergence results for the PAMG approaches considered are displayed in Table 6.5. Note that only 5 levels are performed because of the much smaller size of the system. Namely, the system size is approximately two third of the original size because the highly refined mesh at the electrodes has been cut off.

We observe that the convergence for PAMG(max) is nearly independent of the number of levels employed and that its convergence factor has significantly been improved. This confirms the above assumption that the divergence for the full-level PAMG(max) approach is caused by the critical points.

The convergence factor for PAMG(Ag^+) remains dependent on the number of levels. Nonetheless, the convergence has improved. The level-dependency points to problems with the coarse-grid correction for non-critical points.

The divergence for the PAMG(NO_3^-) approach for the original system is caused by large Péclet numbers on coarse levels. However, its convergence for the modified system is not better than for the one-level iterative solver. The reason being an insufficient coarse-grid correction for points with low migration.

6.2.5   Conclusion

For all PAMG approaches considered, the two scenarios show a similar behavior, namely the convergence of linear systems from late Newton steps (A-30 and B-46) is dependent on the number of levels performed. For the systems from early Newton steps (systems A-5 and B-5), the convergence factors do not differ much from the one-level case. Hence, for these systems the multilevel hierarchy does not pay-off.

The analysis of the late systems A-30 and B-46 indicate that the problems for PAMG(max) are caused by an oscillating smoother at the electrodes on level 5 because of an insufficient coarsening causing large Péclet numbers. The coarsening of PAMG(Ag^+) proceeds differently which seems to be better suited to avoid increasing Péclet numbers on coarser levels. However, comparing the 5-level methods of PAMG(Ag^+) and PAMG(max) the latter shows better convergence factors. The investigations demonstrate that this is caused by an unsuited coarse-grid correction for the points which do not violated the Péclet condition in the case of PAMG(Ag^+).

The approach PAMG(NO_3^-) shows the worst convergence behavior of all approaches considered. This has two reasons. First, its coarse-grid correction for non-critical points is insufficient like in the case of PAMG(Ag^+), and, second, the area of critical points on coarse levels spreads out rather far which, additionally, causes problems for the coarse-grid correction of the critical points. Consequently, the PAMG(NO_3^-) approach shows divergence if performing more than 3 levels for systems A-30 and B-46.

6.3   The Physics-Aware Primary Matrix

We have pointed out that the PAMG approaches investigated in the previous section have a level-dependent convergence. However, the reasons for this differ. PAMG(Ag^+) has a good coarse-grid correction in the direct surrounding of the electrodes, but not in areas with low potential gradient where the Péclet condition is fulfilled. PAMG(max) shows the opposite behavior. Hence, it seems to be a good strategy to combine the coarsening approaches.

Additionally, we observe that the approach PAMG(NO_3^-) suffers from a bad coarse-grid correction especially at non-critical points. However, it is not yet clear if the divergence of the method using 4 and more levels is solely caused by the large Péclet numbers or by a combination of large Péclet numbers and bad coarse-grid correction for the non-critical points. Combining the different coarsening strategies helps to further investigate this.

We derive several PAMG approaches which make use of a primary matrix which combines the primary matrices considered previously. To be more specific, the new approaches use a different primary matrix type near the upper electrode than near the lower one, and a further type for the part of the domain without the area near the electrodes. Since we explicitly use the information about the positions of the electrodes for the construction of the new primary matrix, we call it a physics-aware primary matrix.

We consider approaches named PAMG(A,B,C) where A defines the coarsening near the upper electrode, C the coarsening near the lower electrode and B the coarsening for the remaining points. That is A, B and C are one of max, NO_3^-, or Ag^+. With exception of the primary matrix, the PAMG(A,B,C) approaches use the same components as before.

For example, we define the entries p_{ij} of the primary matrix P for PAMG(Ag^+,max,NO_3^-) to be

$$p_{ij} = \begin{cases} \boldsymbol{A}_{(i,j)}(c(Ag^+), c(Ag^+)) & \text{if } i \in \Omega_1 \\ \|\boldsymbol{A}_{(i,j)}\|_{\max} & \text{if } i \in \Omega_2 \\ \boldsymbol{A}_{(i,j)}(c(NO_3^-), c(NO_3^-)) & \text{if } i \in \Omega_3 \end{cases} , \qquad (6.5)$$

where $\boldsymbol{A}_{(i,j)}$ denotes the point coupling matrix from point i to point j and $\boldsymbol{A}_{(i,j)}(c(Ag^+), c(Ag^+))(\boldsymbol{A}_{(i,j)}(c(NO_3^-), c(NO_3^-)))$ the coupling from unknown $c(Ag^+)$ $(c(NO_3^-))$ at point i to the same unknown at point j. Ω_1 represents the set of points near the upper electrode, Ω_3 the set of points near the lower electrode, and $\Omega_2 = \Omega \backslash (\Omega_1 \cup \Omega_3)$, where Ω is the set of all points. That is, it uses a PAMG(Ag^+)-like coarsening at the upper electrode, a PAMG(NO_3^-)-like coarsening at the lower electrode, and a PAMG(max)-like coarsening for the other points. Note that we have PAMG(A,A,A)=P-AMG(A), with A$\in \{max, Ag^+, NO_3^-\}$.

Remark 6.4. *In this section, the set Ω_1 contains all points within a rectangle which boundary has a distance of 0.005 to the upper electrode, and Ω_3 contains all points within a rectangle which boundary has a distance of 0.005 to the lower electrode, see Figure 6.6.*

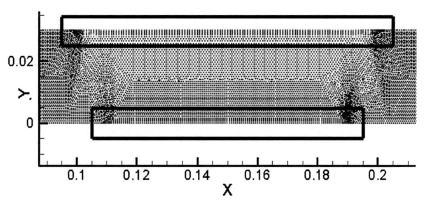

Figure 6.6: Péclet numbers and rectangle domains Ω_1 and Ω_3 around the electrodes for system A-30 (detail). Red points denote a violated Péclet condition.

We begin with a description of the results for the systems from early Newton steps (A-5 and B-5) in Section 6.3.1. Then, we present the results for the linear systems from late Newton steps (A-30 and B-46) in Section 6.3.2. Finally, we summarize the results in Section 6.3.3.

Remark 6.5. *In the following tables, the approaches considered are ordered by their two-level convergence factors and then by their multilevel convergence factor. In this way, approaches which suffer from a strong level-dependency, but with good two-level convergence factor are positioned* first. In Section 6.4, we further implement techniques to reduce the level-dependency of the considered approaches.

Notation 6.6. In the following, the statement "a NO_3^- like primary matrix in Ω_1 leads to the best convergence results" means that $PAMG(NO_3^-,B,C)$ has a better or equal convergence than $PAMG(A,B,C)$ with $A, B, C \in \{NO_3^-, Ag^+, max\}$.

6.3.1　Results for Linear Systems A-5 and B-5

approach			levels						
			1	2	3	4	5	6	7
PAMG(Ag^+	Ag^+	NO_3^-)	0.9	0.49	0.61	0.67	0.71	0.73	-
PAMG(NO_3^-	Ag^+	NO_3^-)	0.9	0.49	0.61	0.66	0.71	0.73	-
PAMG(NO_3^-	max	NO_3^-)	0.9	0.52	0.55	0.56	0.58	0.59	0.65
PAMG(Ag^+	max	NO_3^-)	0.9	0.52	0.55	0.56	0.58	0.6	0.73
PAMG(max	Ag^+	NO_3^-)	0.9	0.53	0.89	0.89	0.89	0.89	-
PAMG(max	max	NO_3^-)	0.9	0.56	0.89	0.89	0.89	0.89	-
PAMG(NO_3^-	Ag^+	max)	0.9	0.68	0.69	0.69	0.64	0.68	0.8
PAMG(Ag^+	Ag^+	max)	0.9	0.69	0.7	0.69	0.64	0.69	0.74
PAMG(NO_3^-	max	max)	0.9	0.71	0.71	0.7	0.68	0.68	0.85
PAMG(Ag^+	max	max)	0.9	0.71	0.71	0.71	0.69	0.69	0.85
PAMG(max	Ag^+	max)	0.9	0.74	0.88	0.88	0.88	0.88	-
PAMG(max	max	max)	0.9	0.77	0.88	0.88	0.88	0.88	-
PAMG(Ag^+	Ag^+	Ag^+)	0.9	0.85	0.88	0.88	0.88	0.88	-
PAMG(NO_3^-	Ag^+	Ag^+)	0.9	0.85	0.88	0.88	0.88	0.88	-
PAMG(Ag^+	max	Ag^+)	0.9	0.86	0.88	0.88	0.88	0.88	-
PAMG(NO_3^-	max	Ag^+)	0.9	0.86	0.88	0.88	0.88	0.88	-
PAMG(Ag^+	NO_3^-	max)	0.9	0.9	0.89	0.89	0.89	0.89	-
PAMG(Ag^+	NO_3^-	NO_3^-)	0.9	0.9	0.89	0.89	0.89	0.89	-
PAMG(NO_3^-	NO_3^-	max)	0.9	0.9	0.89	0.89	0.89	0.89	-
PAMG(NO_3^-	NO_3^-	Ag^+)	0.9	0.9	0.89	0.89	0.89	0.89	-
PAMG(NO_3^-	NO_3^-	NO_3^-)	0.9	0.9	0.89	0.89	0.89	0.89	-
PAMG(Ag^+	NO_3^-	Ag^+)	0.9	0.9	0.9	0.9	0.9	0.9	-
PAMG(max	Ag^+	Ag^+)	0.9	0.92	0.88	0.88	0.88	0.88	-
PAMG(max	max	Ag^+)	0.9	0.93	0.88	0.88	0.88	0.88	-
PAMG(max	NO_3^-	max)	0.9	0.97	0.88	0.88	0.88	0.88	-
PAMG(max	NO_3^-	Ag^+)	0.9	0.97	0.88	0.88	0.88	0.88	-
PAMG(max	NO_3^-	NO_3^-)	0.9	0.97	0.88	0.88	0.88	0.88	-

Table 6.6: Convergence factors for system A-5.

Table 6.6 shows the convergence results for system A-5 for all PAMG approaches considered. The approaches using an PAMG(NO_3^-)-like coarsening for the points in Ω_2, i.e., with low Péclet number, show very bad convergence factors. The approaches using PAMG(Ag^+) and PAMG(max)-like coarsening for these points show the best convergence behavior, except for PAMG(max,Ag^+,Ag^+) and PAMG(max,max,Ag^+).

The differences between the two-level methods of PAMG(A,Ag^+,C) and PAMG(A,max,C) are only marginally ($A, C \in \{NO_3^-, Ag^+, \text{max}\}$). However, using PAMG($Ag^+$)-like coarsening in Ω_2 leads to stronger level-dependency. These results are also in line with the convergence results for system A-30 excluding critical points in Section 6.2.4. Namely, using a non-diffusion like coarsening for non-critical points, the coarse-grid correction leads to convergence problems.

The two-level results indicate that it is important to choose an P-AMG(NO_3^-)-like coarsening in Ω_3. However, considering the multilevel results choosing PAMG(max)-like coarsening in Ω_3 leads to the best convergence factors because of its lower level-dependency.

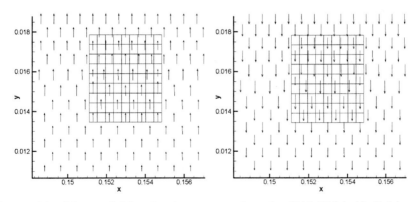

Figure 6.7: Vector fields of primary matrices for PAMG(Ag^+) (left) and PAMG(NO_3^-) (right) for system A-5 (detail with reference grid).

The good performance of the PAMG(NO_3^-)-like coarsening for points in Ω_3 can be explained by the vector field information included in the primary matrices. The vector fields are shown in Figure 6.7. They have been created with AV method described in Section 4.4.2. The vectors of the NO_3^- primary matrix are longer than the ones of Ag^+. The longer the vector, the faster is the movement at a certain point. Hence, the coarsening has less diffusive (regular) character and takes the direction of movement more into account. Note also that only the strength of the vectors and not their direction is important. Hence, assuming the same strength of the vector fields the coarsening would be very similar. The reason for the different strength are the different diffusive properties of the ions. NO_3^- has a significantly larger diffusion coefficient than Ag^+. Hence, it moves faster in the electrolyte than Ag^+.

Considering the two-level methods, the coarsening types do not differ much in Ω_1 for system A-5. However, using PAMG(max)-like coarsening in this domain the convergence factor worsens drastically from 2 to 3 levels. Starting from level 3 its convergence remains constant. The reason are the large Péclet numbers at the upper electrode. In the case of the ion-concentration unknown based coarsening, the Péclet numbers on level 2 are much lower than for PAMG(max)-like coarsening. However, as we already observed, large Péclet numbers on coarse levels nearly vanish for all approaches at the upper electrode. Hence, the convergence stays stable.

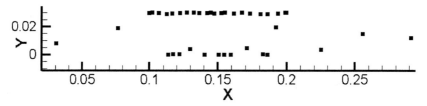

Figure 6.8: Points of level 7 for PAMG(NO_3^-,max,NO_3^-).

We also observe a significant jump in the convergence factors from 6 to 7 levels for the methods which perform 7 levels. This jump can be explained by the distribution of points at the 7th level. Namely, there are nearly any points left in the whole domain besides at the electrodes, see Figure 6.8 which shows the 7th level of PAMG(NO_3^-,max,NO_3^-). Since the smoother does not converge at the 6th and 5th level, this distribution of points leads to problems with the coarse-grid correction.

The distribution is caused by the mixed coarsening strategy. To further explain this, we consider the original approaches. The PAMG(max) approach ends its coarsening after level 6 because there are too few C-Points for level 7. In contrast, PAMG(Ag^+) ends the coarsening on level 6 because of the high density of the coarse level matrix. As a result some of the mixed coarsening strategies have too few C-point on level 7 besides near the electrodes. Hence, the coarse-grid correction is insufficient. To remedy this, the AMG setup has to detect if there are enough points within every region of the domain. This can, for example, be done by providing the coordinates.

approach			levels						
			1	2	3	4	5	6	7
PAMG(NO_3^-	Ag^+	NO_3^-)	0.78	0.43	0.53	0.58	0.62	0.63	
PAMG(Ag^+	Ag^+	NO_3^-)	0.78	0.43	0.53	0.58	0.62	0.63	
PAMG(NO_3^-	max	NO_3^-)	0.78	0.45	0.47	0.49	0.50	0.51	0.57
PAMG(Ag^+	max	NO_3^-)	0.78	0.45	0.47	0.49	0.50	0.52	0.64
PAMG(max	Ag^+	NO_3^-)	0.78	0.53	0.77	0.77	0.77	0.77	
PAMG(max	max	NO_3^-)	0.78	0.56	0.77	0.77	0.77	0.77	
PAMG(NO_3^-	Ag^+	max)	0.78	0.60	0.60	0.60	0.55	0.59	0.69
PAMG(Ag^+	Ag^+	max)	0.78	0.60	0.60	0.60	0.56	0.60	
PAMG(NO_3^-	max	max)	0.78	0.62	0.61	0.61	0.59	0.59	0.74
PAMG(Ag^+	max	max)	0.78	0.62	0.62	0.61	0.60	0.60	0.73

Table 6.7: Convergence factors for system B-5.

Table 6.7 displays the convergence results for system B-5. It only shows the 10 best approaches because the convergence behavior of all approaches considered is very similar to the one of system A-5. However, there is one difference in the character of the results which is caused by the larger current at the upper electrode. The larger current results in worse convergence factors using a PAMG(max)-like coarsening than in the case of a PAMG(Ag^+)- or PAMG(NO_3^-)-like coarsening for points in Ω_1, already, for the two-level approach.

Summarizing, the best approach using 2 levels are PAMG(NO_3^-,Ag^+, NO_3^-) and PAMG(Ag^+,Ag^+,NO_3^-) for systems A-5 and B-5. However, the best multigrid method turns out to be PAMG(NO_3^-,max,NO_3^-) using more than two levels because of the strong level-dependency of PAMG(NO_3^-, Ag^+,NO_3^-) and PAMG(Ag^+,Ag^+,NO_3^-) which is caused by an inappropriate coarsening of the non-critical points. The choice of a PAMG(NO_3^-)-like coarsening at the lower electrode leads to very good results. This is caused by the large diffusion coefficient of NO_3^- which leads to a relatively large migration coefficient causing a very slow coarsening in the direction of migration. The slow coarsening anticipates a spreading of the area with critical points and, hence, a level-dependent convergence behavior.

6.3.2   Results for Linear Systems A-30 and B-46

Table 6.8 shows the convergence results of the approaches considered for system A-30. Similar to systems A-5 and B-5, the approaches using P-AMG(NO_3^-)-like coarsening for points in Ω_2 do not perform well because of

approach			levels						
			1	2	3	4	5	6	7
PAMG(Ag^+	Ag^+	max)	0.96	0.71	0.74	0.75	0.77	0.83	
PAMG(NO_3^-	Ag^+	max)	0.96	0.71	0.74	0.75	0.77	0.83	
PAMG(max	Ag^+	max)	0.96	0.71	0.74	0.75	0.77	0.84	
PAMG(max	max	max)	0.96	0.71	0.74	0.74	0.74	div.	
PAMG(Ag^+	max	max)	0.96	0.71	0.74	0.74	0.74	div.	
PAMG(NO_3^-	max	max)	0.96	0.71	0.74	0.74	0.74	div.	
PAMG(Ag^+	max	Ag^+)	0.96	0.8	0.83	0.84	0.83	0.82	
PAMG(NO_3^-	max	Ag^+)	0.96	0.8	0.83	0.84	0.83	0.82	
PAMG(Ag^+	Ag^+	Ag^+)	0.96	0.8	0.83	0.85	0.86	0.86	
PAMG(NO_3^-	Ag^+	Ag^+)	0.96	0.8	0.83	0.85	0.86	0.86	
PAMG(max	max	Ag^+)	0.96	0.81	0.84	0.85	0.84	0.83	
PAMG(max	Ag^+	Ag^+)	0.96	0.81	0.84	0.85	0.84	0.83	
PAMG(max	Ag^+	NO_3^-)	0.96	0.83	0.88	div.	div.	div.	
PAMG(NO_3^-	Ag^+	NO_3^-)	0.96	0.83	0.88	div.	div.	div.	
PAMG(Ag^+	Ag^+	NO_3^-)	0.96	0.83	0.88	div.	div.	div.	
PAMG(max	max	NO_3^-)	0.96	0.85	0.89	div.	div.	div.	div.
PAMG(Ag^+	max	NO_3^-)	0.96	0.85	0.89	div.	div.	div.	div.
PAMG(NO_3^-	max	NO_3^-)	0.96	0.85	0.89	div.	div.	div.	div.

Table 6.8: Convergence factors for system A-30.

a bad coarse-grid correction for the non-critical points. In the case of system A-30, these approaches show the worst results of all ones considered. Hence, they are not displayed in the table.

For system A-30, the large Péclet numbers spread out very fast independent of the approach. Hence, many large Péclet numbers in Ω_2 arise, see Figure 6.9. Performing a PAMG(max)-like coarsening at the lower electrode leads to slightly better results because the number of large Péclet numbers are reduced faster.

Another consequence of the fast spreading of the large Péclet numbers on coarse levels is that the methods PAMG(*,Ag^+,max) show no divergence in contrast to PAMG(*,max,max) for the full-level approach. The reason being that PAMG(*,Ag^+,max) performs a slightly slower coarsening in Ω_2.

Choosing a PAMG(NO_3^-)-like coarsening in Ω_3 leads to divergence for system A-30 when using more than 3 levels. The divergence can be explained by the distribution of large Péclet numbers which spread faster using PAMG(NO_3^-)-like coarsening at the lower electrode, see Figure 6.9.

Table 6.9 shows the convergence results of the 12 best methods for sys-

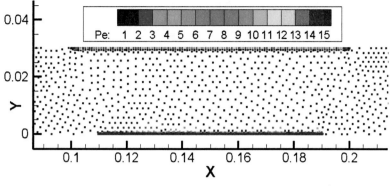

(a) Péclet numbers on level 3 for PAMG(NO_3^-,max,Ag^+).

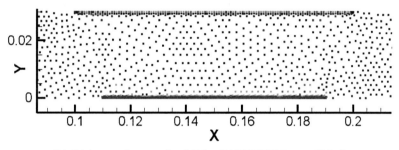

(b) Péclet numbers on level 3 for PAMG(NO_3^-,max,NO_3^-).

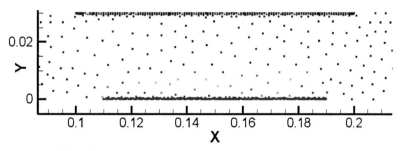

(c) Péclet numbers on level 4 for PAMG(NO_3^-,max,Ag^+).

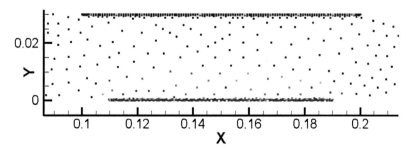

(d) Péclet numbers on level 4 for PAMG(NO_3^-,max,NO_3^-).

Figure 6.9: Péclet numbers on levels 3 and 4 for system A-30.

approach			levels					
			1	2	3	4	5	6
PAMG(max	Ag^+	max)	0.88	0.63	0.65	0.66	0.66	0.75
PAMG(NO_3^-	Ag^+	max)	0.88	0.63	0.65	0.66	0.67	0.75
PAMG(Ag^+	Ag^+	max)	0.88	0.63	0.65	0.66	0.73	0.77
PAMG(Ag^+	max	max)	0.88	0.63	0.65	0.65	0.71	0.78
PAMG(max	max	max)	0.88	0.63	0.65	0.65	0.65	div.
PAMG(NO_3^-	max	max)	0.88	0.63	0.65	0.65	0.65	div.
PAMG(NO_3^-	max	Ag^+)	0.88	0.70	0.73	0.74	0.73	0.76
PAMG(Ag^+	max	Ag^+)	0.88	0.71	0.74	0.75	0.75	0.77
PAMG(NO_3^-	Ag^+	Ag^+)	0.88	0.71	0.74	0.76	0.77	0.77
PAMG(Ag^+	Ag^+	Ag^+)	0.88	0.71	0.74	0.76	0.77	0.78
PAMG(max	max	Ag^+)	0.88	0.71	0.74	0.75	0.74	0.74
PAMG(max	Ag^+	Ag^+)	0.88	0.72	0.75	0.76	0.77	0.77

Table 6.9: Convergence factors for system B-46.

tem B-46 which are very similar to the ones of system A-30. The convergence results of the PAMG(NO_3^-)-like coarsening at the lower electrode are not included in the table, alike as in the case of system A-30 these methods diverge starting from level 3. We assume that the methods using PAMG(NO_3^-)-like coarsening at the lower electrode would perform significantly better if the problems caused by large Péclet numbers on coarse levels are lowered.

In contrast to system B-5, the convergence factors of the two-level methods are not sensitive to the coarsening approach at the upper electrode. This is most likely caused by the highly spread out distribution of large Péclet numbers already on level 2 at the lower electrode, which overcompensates the effect of a ion-concentration unknown based coarsening.

The best approach for systems A-30 and B-46 is PAMG(*,Ag^+,max) because it uses a migration dependent coarsening between the electrodes, where large Péclet numbers on coarse levels develop. It is significantly better than PAMG(*,NO_3^-,max) which does nearly ignore the diffusion because of its strong migration field. Additionally, using PAMG(max)-like coarsening at the lower electrode helps to reduce the amount of large Péclet numbers in this area resulting in a lower level-dependency.

6.3.3   Summary of the Results

We have demonstrated that physics-aware primary matrices are able to out-perform state-of-the-art approaches significantly in terms of convergence factors. The physics-aware primary matrices introduced take the positions of the electrodes into account to stress the importance of the migration near the electrodes and to stress the diffusion where the potential gradients are low. This strategy is especially beneficial for systems from early Newton steps like systems A-5 and B-5. For these systems, the approximation of the potential is far away from the solution, which leads to an underrepresentation of the migration near the electrodes using a norm-based primary matrix.

However, some of the methods using a physic-aware primary matrix perform an additional level. This level does harm their convergence factor significantly. The reason for the significant worsening of the convergence is that the distribution of points on coarse levels leads to an insufficient coarse-grid correction. We show that this can be remedied by tuning the physics-aware primary matrix with respect to the geometry and ion-system considered in Section 6.5.

Considering the systems from late Newton steps (A-30 and B-46), using the physics-aware primary matrix does not show divergence like the PAMG(max) approach which is the best state-of-the-art approach for early Newton steps. The divergence of PAMG(max) is caused by the distribution of large Péclet numbers on the coarse levels. This property leads to a strong level-dependency for all approaches considered. Hence, we investigate how we can deal with a locally violated Péclet condition in the case of the migration-diffusion system to prevent the level-dependency of the approaches.

6.4   PAMG for a Locally Violated Péclet Condition

For the scalar case, we have tested several strategies to prevent increasing Péclet numbers on the coarse levels, see Section 5.1. The strategies considered have been

- considering critical variables only within the fine level smoothing process, that is, we do not restrict these variables to the coarse levels (SMO),

- solving a subsystems containing the critical variables with a direct solver and coupling it to the other variables in an alternating Schwarz method fashion (AS), and

- forcing critical variables to exist on all levels (FC).

We generalize these strategies for the system case. Since we consider a strongly coupled PDE system and apply a point-based approach, all variables at a point have to be treated alike. The heuristic Péclet number for the MITReM defined in Section 5.2 is already defined point-wise, hence, the transfer of these methods to the system case is straightforward.

We investigate if these strategies improve the convergence behavior of the PAMG approaches considered and compare the quality of the results to the scalar case of the convection-diffusion equation. We start with a description of the methods using forced F-points, i.e. SMO and AS, in Section 6.4.1 and, then, present the results for the method using forced C-points, i.e. FC, in Section 6.4.2.

Remark 6.7. *We use the rule that a point becomes a forced C- (or F-) point whenever the heuristic Péclet number at the point is greater than 2, i.e. it violates the Péclet condition.*

6.4.1   Using Forced F-Points

SMO Strategy

scenario	linear	levels					
	system	1	2	3	4	5	6
A	5	0.90	0.85	0.85	0.85	0.85	0.85
	30	0.96	0.94	0.94	0.94	0.94	0.94
B	5	0.78	0.74	0.74	0.74	0.74	0.74
	46	0.88	0.83	0.83	0.83	0.83	0.83

Table 6.10: Convergence factors for PAMG(max)-SMO(0).

We consider the method PAMG(max)-SMO(0). In Table 6.10, we see that the convergence of the PAMG(max)-SMO(0) approach is independent of the number of levels employed. However, the convergence factor is not much better than in the one-level case. Hence, the multigrid hierarchy does not pay-off. Similar to the scalar case, using an overlap has no impact on the convergence factors.

Comparing the convergence results to PAMG(max), the convergence for the systems from early Newton steps has improved. However, the convergence is worse for late Newton steps with exception of the full-level method where PAMG(max) diverged.

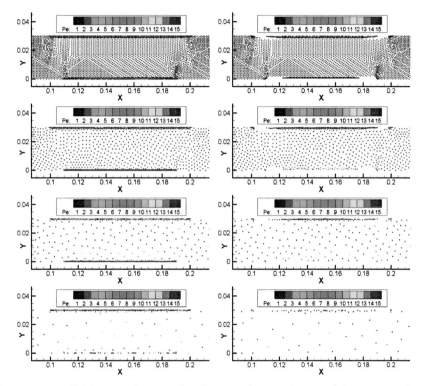

Figure 6.10: Péclet numbers on level 2 to 5 (top to bottom) for system B-46 for PAMG(max) (left) and PAMG(max)-SMO (right) (detail).

Figure 6.10 shows that the number of points which violate the Péclet condition on the coarse levels of PAMG(max)-SMO is lower compared to the original approach. Additionally, the BGS smoother converges for all coarse level systems arising for PAMG(max)-SMO.

In contrast to some scalar examples BGS converges on the fine level for the problems considered. Since PAMG(max) converges well for the non-critical points, cf. Section 6.2.4 and only few new critical points arise on coarse levels, PAMG(max)-SMO(0) converges.

Figure 6.11 shows coarsening strategies of three PAMG approaches. PAMG(Ag^+)-SMO nearly avoids large Péclet numbers on the coarse levels, and PAMG(NO_3^-,max,NO_3^-)-SMO's number of critical points is between

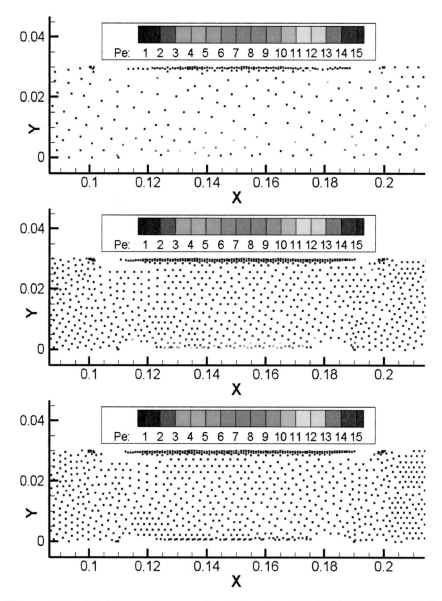

Figure 6.11: Péclet numbers on level 3 using PAMG(NO_3^-,max,NO_3^-)-SMO, PAMG(max)-SMO, and PAMG(Ag^+)-SMO (top to bottom) for system B-46 (detail).

PAMG(max)-SMO and PAMG(Ag^+)-SMO. However, the convergence factors for all three methods are exactly the same.

The results indicate that SMO's convergence is restricted by the BGS smoother's convergence for the critical points on the first level. This is additionally underlined by the observation that using an overlap for the methods considered does not improve the convergence. Hence, convergence results for the other PAMG approaches considered are similar and, therefore, not displayed.

AS Strategy

scenario	linear system	\multicolumn{6}{c}{levels}
		1	2	3	4	5	6
A	5	0.89	0.90	0.90	0.90	0.90	090
	30	0.96	0.98	0.98	0.98	0.98	0.98
B	5	0.77	0.78	0.78	0.78	0.78	0.78
	46	0.97	0.98	0.98	0.98	0.98	0.98

Table 6.11: Convergence factors for PAMG(max)-AS(0).

Table 6.11 shows the convergence results for PAMG(max)-AS(0). The convergence factors are independent of the number of levels employed. However, the convergence is even slightly worse than for the one-level iterative solver. Hence, the multigrid hierarchy does not pay-off.

Again, it seems that the outer convergence restricts the overall convergence factor of this method. This is underlined by the fact that the convergence factors of PAMG(Ag^+)-AS(0) and PAMG(NO_3^-,max,NO_3^-)-AS(0) are exactly the same as the ones for PAMG(max)-AS. Another indicator is the good convergence factors of PAMG(Ag^+) and PAMG(max) applied to system A-30 excluding the critical areas which we have observed in Section 6.2.4. Consequently, only results for PAMG(max)-AS(0) are displayed.

An overlap does not improve the convergence factors for PAMG(max)-AS(0) as we have observed in the scalar case. The effects are comparable to the ones observed in the scalar case when working with jumping coefficients. It seems that the migration coefficient which gradient is very steep near the electrodes causes this behavior.

6.4.2   Using Forced C-Points

We investigate PAMG approaches using forced C-points. In the case of forced C-points, all critical points are present at the coarsest level. Hence, their correction is computed exactly at the coarsest level. Furthermore, they are coupled to the other points on all levels which usually improves the coarse-grid correction for the point in the direct neighborhood.

As before, the behavior of the approaches for the systems A-5 and B-5 do not differ much, also the convergence results for systems A-30 and B-46 are very similar. Hence, we first analyze the systems from early Newton steps and then the latter ones for PAMG using forced C-points.

Results for Linear Systems A-5 and B-5

Table 6.12 shows the convergence results for the PAMG approaches considered using forced C-points for system A-5. Note that the PAMG-FC(0) approaches do only perform 5 levels because of a too slow coarsening between levels 5 and 6 which is caused by the large number of forced C-points.

As in the case without forced C-points, the choice of the coarsening-type for the points in Ω_2 has the largest impact on the convergence factor. For the case of PAMG using forced C-points, PAMG(max)-like coarsening is significantly better than PAMG(Ag^+)-like coarsening in Ω_2. This is in contrast to the convergence results for PAMG without using forced C-points, where the results of PAMG(max)-like and PAMG(Ag^+)-like coarsening in Ω_2 are similar, cf. Section 6.3.

The reason for the different behavior is that less large Péclet numbers arise in Ω_2 on coarse levels, because forced C-points are used. Therefore, the improvement of the Péclet numbers by an PAMG(Ag^+)-like coarsening is overcompensated by the disadvantage of a level-dependent convergence factor for the non-critical points.

Similar to the case without forced C-points, the PAMG(NO_3^-)-like coarsening for non-critical points causes a very bad convergence. Hence, approaches using PAMG(NO_3^-)-like coarsening for this area perform worse than PAMG(Ag^+)-like or maximum-norm based coarsening-types. This is clear since using forced C-points does not help to improve the coarse-grid correction for non-critical points.

At the upper electrode, the PAMG(Ag^+) and PAMG(NO_3^-)-like coarsening-types perform best. Using PAMG(max)-like coarsening in Ω_1 causes a drastic drop in the convergence factor between 2 and 3 levels which we have already observed for the PAMG approaches which do not use forced

approach			1	2	3	4	5
PAMG(Ag^+	max	NO_3^-)-FC(0)	0.9	0.56	0.56	0.57	0.57
PAMG(NO_3^-	max	NO_3^-)-FC(0)	0.9	0.56	0.56	0.57	0.57
PAMG(Ag^+	max	max)-FC(0)	0.9	0.56	0.57	0.57	0.57
PAMG(Ag^+	max	Ag^+)-FC(0)	0.9	0.56	0.56	0.57	0.57
PAMG(NO_3^-	max	max)-FC(0)	0.9	0.56	0.57	0.57	0.57
PAMG(NO_3^-	max	Ag^+)-FC(0)	0.9	0.56	0.56	0.57	0.59
PAMG(Ag^+	Ag^+	max)-FC(0)	0.9	0.56	0.64	0.69	0.7
PAMG(Ag^+	Ag^+	Ag^+)-FC(0)	0.9	0.56	0.64	0.69	0.7
PAMG(Ag^+	Ag^+	NO_3^-)-FC(0)	0.9	0.56	0.64	0.69	0.7
PAMG(NO_3^-	Ag^+	max)-FC(0)	0.9	0.56	0.64	0.69	0.7
PAMG(NO_3^-	Ag^+	Ag^+)-FC(0)	0.9	0.56	0.64	0.69	0.7
PAMG(NO_3^-	Ag^+	NO_3^-)-FC(0)	0.9	0.56	0.64	0.69	0.7
PAMG(max	max	NO_3^-)-FC(0)	0.9	0.6	0.87	0.86	0.86
PAMG(max	Ag^+	max)-FC(0)	0.9	0.6	0.87	0.86	0.86
PAMG(max	Ag^+	Ag^+)-FC(0)	0.9	0.6	0.87	0.86	0.86
PAMG(max	Ag^+	NO_3^-)-FC(0)	0.9	0.6	0.87	0.86	0.86
PAMG(max	max	max)-FC(0)	0.9	0.6	0.87	0.86	0.86
PAMG(max	max	Ag^+)-FC(0)	0.9	0.6	0.87	0.86	0.86
PAMG(NO_3^-	NO_3^-	Ag^+)-FC(0)	0.9	0.9	0.9	0.9	0.92
PAMG(Ag^+	NO_3^-	Ag^+)-FC(0)	0.9	0.9	0.9	0.9	0.94
PAMG(NO_3^-	NO_3^-	max)-FC(0)	0.9	0.9	0.9	0.9	0.96
PAMG(Ag^+	NO_3^-	max)-FC(0)	0.9	0.9	0.9	0.9	0.99
PAMG(Ag^+	NO_3^-	NO_3^-)-FC(0)	0.9	0.9	0.9	0.9	div.
PAMG(NO_3^-	NO_3^-	NO_3^-)-FC(0)	0.9	0.9	0.9	0.9	div.
PAMG(max	NO_3^-	Ag^+)-FC(0)	0.9	0.96	0.88	0.88	0.92
PAMG(max	NO_3^-	max)-FC(0)	0.9	0.96	0.88	0.88	0.97
PAMG(max	NO_3^-	NO_3^-)-FC(0)	0.9	0.96	0.88	0.88	div.

Table 6.12: Convergence factors for system A-5 using forced C-points.

C-points.

The reason is that the migration at the upper electrode is very low for systems from early Newton steps, like system A-5. Hence, a maximum-norm based primary matrix causes a diffusion-driven coarsening at the upper electrode, cf. Figure 6.2. The diffusion-driven coarsening leads to more points violating the Péclet condition on coarser levels than in the case of an ion-concentration unknown based coarsening. The reason being that the distances between the coarse level points increase faster. This also explains the drop in the convergence factor between 2 and 3 levels for PAMG(max,B,*) with $B \in \{Ag^+, max\}$.

At the lower electrode, all coarsening-types show nearly the same convergence factors. In contrast to the upper electrode, the Péclet numbers are larger and captured well by using forced C-points. Hence, less additional critical points develop compared to Ω_1.

approach			levels				
			1	2	3	4	5
PAMG(Ag^+	max	NO_3^-)-FC(0)	0.78	0.48	0.49	0.49	0.50
PAMG(Ag^+	max	max)-FC(0)	0.78	0.48	0.49	0.49	0.49
PAMG(NO_3^-	max	max)-FC(0)	0.78	0.48	0.49	0.49	0.50
PAMG(NO_3^-	max	NO_3^-)-FC(0)	0.78	0.48	0.49	0.49	0.50
PAMG(Ag^+	max	Ag^+)-FC(0)	0.78	0.48	0.49	0.49	0.50
PAMG(NO_3^-	max	Ag^+)-FC(0)	0.78	0.48	0.49	0.49	0.51
PAMG(NO_3^-	Ag^+	NO_3^-)-FC(0)	0.78	0.49	0.55	0.60	0.61
PAMG(Ag^+	Ag^+	Ag^+)-FC(0)	0.78	0.49	0.55	0.60	0.61
PAMG(Ag^+	Ag^+	NO_3^-)-FC(0)	0.78	0.49	0.55	0.60	0.61
PAMG(NO_3^-	Ag^+	max)-FC(0)	0.78	0.49	0.55	0.60	0.61
PAMG(NO_3^-	Ag^+	Ag^+)-FC(0)	0.78	0.49	0.55	0.60	0.61
PAMG(Ag^+	Ag^+	max)-FC(0)	0.78	0.49	0.55	0.60	0.61
PAMG(max	max	Ag^+)-FC(0)	0.78	0.60	0.75	0.75	0.75
PAMG(max	max	NO_3^-)-FC(0)	0.78	0.60	0.75	0.75	0.75
PAMG(max	max	max)-FC(0)	0.78	0.60	0.75	0.75	0.75
PAMG(max	Ag^+	max)-FC(0)	0.78	0.60	0.75	0.75	0.75
PAMG(max	Ag^+	Ag^+)-FC(0)	0.78	0.60	0.75	0.75	0.75
PAMG(max	Ag^+	NO_3^-)-FC(0)	0.78	0.60	0.75	0.75	0.75

Table 6.13: Convergence factors for system B-5 using forced C-points.

Table 6.13 shows the convergence results for the PAMG approaches considered using forced C-points for system B-5. The convergence results are very similar to the system A-5 system. Hence, we do not show the results using PAMG(NO_3^-)-like coarsening for the non-critical points, which shows very bad convergence results.

The only significant difference of system B-5 compared to system A-5 is that the methods using PAMG(Ag^+) or PAMG(NO_3^-)-like coarsening show a better convergence because of the larger current at the upper electrode. Using PAMG(max)-like coarsening in this part of the domain performs nearly alike as in the case of system A-5. The only difference is that the drop in the convergence between 2 and 3 levels is not that drastic for this method . This effect can be explained by the larger Péclet numbers at the upper electrode for system B-5. To be more specific, in the case of PAMG(max)-coarsening in Ω_1, the area violating the Péclet condition spreads out faster than in the case where an ion-concentration unknown based coarsening is used.

We compare the convergence results of the state-of-the-art approaches with its counterparts using forced C-points. The convergence factors for systems A-5 and B-5 for PAMG(Ag^+)-FC(0) are improved compared to the original approach PAMG(Ag^+). The reason is the better coarse-grid correction at the electrodes because of the use of forced C-points and, furthermore, the choice of the primary matrix. However, the convergence factor of PAMG(Ag^+)-FC(0) is still level-dependent. The reason is the bad coarse-grid correction for the areas with low migration, which we have already seen for the original approach and is also verified by the performance of PAMG(Ag^+,max,Ag^+)-FC(0).

Using forced C-points does not improve PAMG(NO_3^-) since its problems are not caused by large Péclet numbers on coarse levels, but by an insufficient coarse-grid correction of the non-critical points. The reason is the coarsening which takes the migration direction into account for all points even if this effect is so weak that it can be neglected.

Summarizing, PAMG(NO_3^-,max,NO_3^-)-FC(0) still shows the best convergence factors for systems A-5 and B-5. In particular, the convergence results of the two-level method are now the best for these systems because of the usage of forced C-points. Furthermore, its convergence factors are independent of the number of levels performed.

Results for Linear Systems A-30 an B-46

Table 6.14 shows the convergence results for the PAMG approaches considered using forced C-points for system A-30. Similar to the case of systems A-5 and B-5, the choice of the coarsening for the non-critical points has the

approach			1	2	3	4	5
PAMG(Ag^+	max	NO_3^-)-FC(0)	0.96	0.58	0.61	0.62	0.63
PAMG(NO_3^-	max	NO_3^-)-FC(0)	0.96	0.58	0.61	0.62	0.63
PAMG(max	max	NO_3^-)-FC(0)	0.96	0.59	0.61	0.62	0.63
PAMG(max	max	Ag^+)-FC(0)	0.96	0.59	0.62	0.63	0.65
PAMG(Ag^+	max	Ag^+)-FC(0)	0.96	0.59	0.62	0.63	0.65
PAMG(NO_3^-	max	Ag^+)-FC(0)	0.96	0.59	0.62	0.63	0.65
PAMG(max	max	max)-FC(0)	0.96	0.6	0.61	0.61	0.61
PAMG(Ag^+	max	max)-FC(0)	0.96	0.6	0.61	0.61	0.61
PAMG(NO_3^-	max	max)-FC(0)	0.96	0.6	0.61	0.61	0.61
PAMG(max	Ag^+	NO_3^-)-FC(0)	0.96	0.61	0.7	0.76	0.79
PAMG(Ag^+	Ag^+	NO_3^-)-FC(0)	0.96	0.61	0.7	0.76	0.79
PAMG(NO_3^-	Ag^+	NO_3^-)-FC(0)	0.96	0.61	0.7	0.76	0.79
PAMG(max	Ag^+	max)-FC(0)	0.96	0.61	0.7	0.78	0.81
PAMG(Ag^+	Ag^+	max)-FC(0)	0.96	0.61	0.7	0.78	0.81
PAMG(NO_3^-	Ag^+	max)-FC(0)	0.96	0.61	0.7	0.78	0.81
PAMG(max	Ag^+	Ag^+)-FC(0)	0.96	0.61	0.72	0.79	0.84
PAMG(Ag^+	Ag^+	Ag^+)-FC(0)	0.96	0.61	0.72	0.8	0.84
PAMG(NO_3^-	Ag^+	Ag^+)-FC(0)	0.96	0.61	0.72	0.79	0.84
PAMG(max	NO_3^-	Ag^+)-FC(0)	0.96	0.96	0.96	0.96	0.96
PAMG(Ag^+	NO_3^-	Ag^+)-FC(0)	0.96	0.96	0.96	0.96	0.96
PAMG(NO_3^-	NO_3^-	Ag^+)-FC(0)	0.96	0.96	0.96	0.96	0.96
PAMG(max	NO_3^-	max)-FC(0)	0.96	0.96	0.96	div.	div.
PAMG(max	NO_3^-	NO_3^-)-FC(0)	0.96	0.96	0.96	div.	div.
PAMG(Ag^+	NO_3^-	max)-FC(0)	0.96	0.96	0.96	div.	div.
PAMG(Ag^+	NO_3^-	NO_3^-)-FC(0)	0.96	0.96	0.96	div.	div.
PAMG(NO_3^-	NO_3^-	max)-FC(0)	0.96	0.96	0.96	div.	div.
PAMG(NO_3^-	NO_3^-	NO_3^-)-FC(0)	0.96	0.96	0.96	div.	div.

Table 6.14: Convergence factors for system A-30 using forced C-points.

largest influence on the convergence factor. A PAMG(max)-like coarsening in Ω_2 is the best choice within the coarsening-types considered for the same reasons as in the case of the early systems A-5 and B-5. The reasons are the absence of major problems in this area caused by violated Péclet conditions on coarse levels, and the coarsening for the points based on $c(Ag^+)$ or $c(NO_3^-)$ couplings which is unsuited in this part of the domain.

The choice of the coarsening at the upper electrode has no influence on the convergence factor. The influence of the choice of the coarsening on the convergence is very low at the lower electrodes. To be more specific, using PAMG(max)-like coarsening in Ω_3 seems to fully remove the level-dependency, using the other two methods shows nearly no level-dependency.

The reason is that, in contrast to the behavior for systems A-5 and B-5, the smoother does not diverge in the part of the domain near the upper electrode for the systems from late Newton steps. Figure 6.12 shows this behavior for systems A-5 and A-30 exemplary. Hence, the choice of the coarsening strategy at the upper electrode has no influence on the convergence, which we have already seen for system A-30.

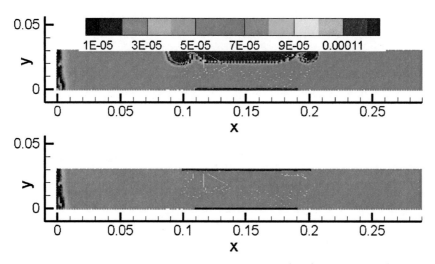

Figure 6.12: Error on level 2 for systems A-5 (top) and A-30 (bottom) after 9 BGS steps using random first guess and zero right hand side for PAMG(NO_3^-,max,NO_3^-).

These observations are in contrast to the findings for the systems from early Newton steps A-5 and B-5. The reason is that in the case of system A-30 the Péclet numbers on the first level are so large that more points are

forced to coarser levels, so that additional points with large Péclet numbers
on coarse levels are nearly avoided.

The slight level-dependency of the ion-concentration unknown based
coarsening at the electrodes is caused by the non-critical points in Ω_3 which
need mainly diffusion-driven coarsening. The effects are comparable to the
behavior of non-critical points in Ω_2. Another reason for this effect is that
the maximum-norm based primary matrix does represent the migration sufficiently well for this system because it is more dominant than for the early
systems, cf. Figure 6.2.

approach			levels			
			1	2	3	4
PAMG(Ag^+	max	NO_3^-)-FC(0)	0.88	0.58	0.62	0.62
PAMG(max	max	NO_3^-)-FC(0)	0.88	0.58	0.62	0.62
PAMG(NO_3^-	max	NO_3^-)-FC(0)	0.88	0.59	0.62	0.62
PAMG(Ag^+	max	Ag^+)-FC(0)	0.88	0.60	0.62	0.63
PAMG(max	max	Ag^+)-FC(0)	0.88	0.60	0.62	0.63
PAMG(NO_3^-	max	Ag^+)-FC(0)	0.88	0.60	0.62	0.63
PAMG(max	max	max)-FC(0)	0.88	0.60	0.61	0.61
PAMG(Ag^+	max	max)-FC(0)	0.88	0.60	0.61	0.61
PAMG(NO_3^-	max	max)-FC(0)	0.88	0.60	0.61	0.61

Table 6.15: Convergence factors for system B-46 using forced C-points.

Table 6.15 shows the convergence results using forced C-points for system B-46. The convergence results are very similar to system A-30, hence,
we only show the 9 best methods. Overall, the methods displayed in the
table hardly show any level-dependency. Also, the difference between the
convergence factors of the approaches is only marginally. Note that only 4
levels are performed because of the large number of forced C-points caused
by the large potential difference between the electrodes.

PAMG(max)-FC(0)'s convergence factor for the linear systems from the
late Newton steps is independent of the number of levels and has improved
compared to the original approach without forced C-points. In contrast
to system A-30, PAMG(Ag^+,max,Ag^+)-FC(0) shows the same convergence
factors as PAMG(max)-FC(0). This is caused by the larger migration at the
upper electrode, which results in less non-critical points. Hence, the ion-
concentration unknown based coarsening is not level-dependent anymore in
this part of the domain.

6.5   The Final Physics-Aware PAMG Approach

We summarize the numerical effects occurring when applying PAMG to the migration-diffusion system considered. Based on the observation, we derive a physics-aware PAMG approach. We consider two scenarios A and B which use different potential differences between the electrodes. For the analysis, we extracted single linear systems from the 5th Newton step (called A-5 and B-5) as a representative for early Newton steps, and the 30th in the case of Scenario A (A-30) and 46th in the case of Scenario B (B-46) as representative for late Newton steps, see Section 6.1 for more details on the systems considered.

Two aspects make the migration-diffusion problem considered hard to solve. The fist aspect is that the system is non-linear and, furthermore, the initial guess for the potential gradient is zero. Hence, the migration in the linear systems from early Newton steps is low. This causes an insufficient representation of the migration using PAMG(max) which is based on a maximum-norm based primary matrix.

The second aspect is that state-of-the-art PAMG becomes level-dependent if the Péclet condition on the fine level is locally strongly violated. This is especially observed for systems from late Newton steps because the potential gradient's steepness increases until the overall solution is reached.

The first aspect is handled by applying a physics-aware primary matrix, which takes the different physical regimes of the domain into account. The second aspect is handled by using forced C-points which means that all critical points are forced to the coarsest level. In the following, we summarize our findings for systems from early and late Newton steps (Section 6.5.1 and Section 6.5.2, respectively) and give some final conclusions in Section 6.5.3.

Remark 6.8. *We have also investigated whether using forced F-points helps to improve the convergence of the PAMG approaches. In the scalar case, we observe that these methods improve the convergence considering a steady convection coefficient. However, the investigations show that these methods do not improve the convergence factor in the case considered here. Their convergence factor is similar to the one of the one-level iterative solver. It seems that their convergence factor is bounded by the convergence of the smoother for the critical points in case of SMO or by the outer convergence in the case of AS. The reason is that the gradient of the potential, i.e. migration coefficient, is too steep towards the electrodes to provide a good outer convergence of the method in the case of the migration-diffusion model considered, see Section 6.4.1 for more details.*

6.5.1 Linear Systems from Early Newton Steps

In early Newton steps, the approximation of the potential unknown is far away from the final solution. This leads to an insufficient coarsening using the maximum-norm based primary matrix, especially, at the upper electrode. Using an ion-concentration as primary unknown instead of the maximum-norm based primary matrix does represent the migration well. However, this approach suffers from a bad coarse-grid correction for points far away from the electrode which have nearly no migration.

Therefore, we suggest a splitting of the domain into three parts. One part contains the points near the upper electrode, one part contains the points near the lower electrode, and one part contains the remaining points. With this partitioning of the domain, a different coarsening can be chosen in each sub-domain. The approaches introduced are called PAMG(A,B,C) where A and C denote the type of the primary matrix in the part of the domain near the upper and lower electrode, respectively, and B denotes the type of the primary matrix in the remaining part.

The combination PAMG(ion,max,ion) gives the best results for early Newton steps. The approach uses a maximum-norm based primary matrix for points away from the electrode. Furthermore, an ion-concentration primary unknown (either $c(NO_3^-)$ or $c(Ag^+)$) is used in the part of the domain near the electrodes in order to represent the migration in this area for early systems well. Note that the use of the NO_3^- ion-concentration leads to a slightly better approach than the use of the Ag^+ ion-concentration. The reason is that NO_3^- has a larger diffusion coefficient which leads to a larger mobility of the ion and, hence, to a larger migration coefficient.

approach			d	levels		
				2	3	full
A-5						
PAMG(NO_3^-	max	NO_3^-)	0.0005	0.52	0.55	0.65 (7)
PAMG(NO_3^-	max	NO_3^-)	0.00002	0.48	0.54	0.54 (7)
B-5						
PAMG(NO_3^-	max	NO_3^-)	0.0005	0.45	0.47	0.57 (7)
PAMG(NO_3^-	max	NO_3^-)	0.00002	0.42	0.47	0.47 (7)

Table 6.16: Convergence factors for linear systems A-5 and B-5. Numbers in brackets denote the number of levels preformed.

Originally, we have used rectangles which boundaries have a distance d of $d = 0.0005$ to the electrodes in order to separate the domain into three parts, cf. Figure 6.6. This splitting has been chosen to simplify the numerical analysis. However, it is not optimal in terms of convergence factor. Table 6.16 shows exemplary the convergence results for $PAMG(NO_3^-,\text{max},NO_3^-)$ with the distance d=0.0005 originally used, and the distance d=0.00002 which shows the best results. The optimal distance depends on the physical properties and discretization. Hence, it might be different if considering other geometries and/or refinements. The new distance equals approximately eight grid points in horizontal direction.

The new partitioning does not show the drastic drop in the convergence factor from 6 to 7 levels we have seen for systems A-5 and B-5. The drop for $d = 0.0005$ is caused by the fact that PAMG is not able to detect that there are too few points in the inner of the domain to compute a good coarse-grid correction for level 6. With the new size the distribution of points, especially, on the coarsest level is better. Hence, the drop of the convergence between 6 and 7 levels vanishes.

approach			d	levels		
				2	3	full
A-5						
$PAMG(NO_3^-$	max	$NO_3^-)$-FC(0)	0.0005	0.56	0.56	0.57 (5)
$PAMG(NO_3^-$	max	$NO_3^-)$-FC(0)	0.00002	0.56	0.57	0.57 (5)
$PAMG(NO_3^-$	max	$NO_3^-)$-FC(1)	0.00002	0.59	0.57	0.58 (5)
B-5						
$PAMG(NO_3^-$	max	$NO_3^-)$-FC(0)	0.0005	0.48	0.49	0.50 (5)
$PAMG(NO_3^-$	max	$NO_3^-)$-FC(0)	0.00002	0.48	0.49	0.50 (5)
$PAMG(NO_3^-$	max	$NO_3^-)$-FC(1)	0.00002	0.59	0.51	0.50 (5)

Table 6.17: Convergence factors for linear systems A-5 and B-5 using forced C-points. Numbers in brackets denote the number of levels preformed.

We have also investigated whether using forced C-points helps to improve the convergence for the systems from early Newton steps. To be more specific, we have forced points which violate the Péclet condition to become C-points (FC(0)). In the scalar case, we have observed that, additionally, forcing the direct neighbors of points violating the Péclet condition to the coarsest level often helps to further improve the results (FC(1)).

Table 6.17 shows the convergence results for $PAMG(NO_3^-,\text{max},NO_3^-)$ using forced C-points. Generally, using the improved choice of d, the use of forced C-points does not improve the convergence factor of PAMG(ion,max, ion) in the case of systems from early Newton steps. This is due to the fact that this approach does not suffer from problems caused by large Péclet numbers on coarse levels because the critical points for early Newton steps do not strongly violate the Péclet condition. In contrast, the number of large Péclet numbers even lowers on coarse levels. Hence, the use of forced C-points leads to slightly worse convergence factors than in the case without forced C-points.

Overall, the best approach for linear systems from early Newton steps is $PAMG(NO_3^-,max,NO_3^-)$. The use of forced C-points does not improve the method because the points on the fine level only slightly violate the Péclet condition, if at all.

6.5.2   Linear Systems from Late Newton Steps

We already pointed out that the large Péclet numbers are caused by the mesh which has many small angles near the electrode because it is highly refined in y-direction and not simultaneously in x-direction. In the model problem considered, the Péclet numbers increase the better we approximate the final solution, because the initial potential gradient is zero and its steepness increases until the solution is reached. Hence, especially for the systems from late Newton steps, the Péclet numbers are large.

For the scalar model problem, we have observed that the convergence factors will get worse the more levels we introduce if the areas with points which violate the Péclet number exceed a certain size. This is also seen in the case of the migration-diffusion model problem. We observe that the distribution of large Péclet numbers spreads out on coarser levels which causes a worsening of the convergence factor with rising number of levels employed.

The $PAMG(max,Ag^+,max)$ approach shows the best full-level results for systems from late Newton steps because the coarsening based on the ion-concentration coupling takes the migration-direction into account which helps to reduce the spreading of the areas with large Péclet numbers.

The use of forced C-points prevents increasing Péclet numbers and a spreading of the area with large Péclet numbers on coarse levels. Hence, using forced C-points, the PAMG approaches using a maximum-norm or ion-concentration unknown based primary matrix at the upper and lower electrode, and a maximum-norm based primary matrix for the points in Ω_2 show the best results for late Newton steps.

The new distance d also improves the convergence for the late systems, see Table 6.18. The convergence factors of $PAMG(NO_3^-,max,NO_3^-)$ and

approach			d	levels		
				2	3	full
A-30						
PAMG(NO_3^-	max	NO_3^-)	0.0005	0.85	0.89	div. (6)
PAMG(NO_3^-	max	NO_3^-)-FC(0)	0.0005	0.58	0.61	0.63 (5)
PAMG(NO_3^-	max	NO_3^-)	0.00002	0.83	0.87	div. (6)
PAMG(NO_3^-	max	NO_3^-)-FC(0)	0.00002	0.59	0.59	0.61 (5)
PAMG(NO_3^-	max	NO_3^-)-FC(1)	0.00002	0.49	0.50	0.53 (5)
B-46						
PAMG(NO_3^-	max	NO_3^-)	0.0005	0.73	0.78	div. (6)
PAMG(NO_3^-	max	NO_3^-)-FC(0)	0.0005	0.59	0.62	0.62 (4)
PAMG(NO_3^-	max	NO_3^-)	0.00002	0.73	0.76	div. (6)
PAMG(NO_3^-	max	NO_3^-)-FC(0)	0.00002	0.60	0.59	0.61 (4)
PAMG(NO_3^-	max	NO_3^-)-FC(1)	0.00002	0.50	0.50	0.52 (4)

Table 6.18: Convergence factors for systems A-30 and B-46.

PAMG(NO_3^-,max,NO_3^-)-FC(0) are now even marginally better than the ones of PAMG(max) and PAMG(max)-FC(0), respectively. Table 6.18 also shows that it additionally pays off to force the direct neighbors on the coarse level. Forcing the direct neighbors nearly remedies the spreading of large Péclet numbers on coarse levels because it slows down the development of newly arising large Péclet numbers.

6.5.3   Conclusion

We have demonstrated that the method PAMG(NO_3^-,max,NO_3^-) provides the best results for systems from early Newton steps, where the current approximation is far away from the solution. However, with better accuracy of the approximation convergence problems arise caused by points violating the Péclet condition. This can be remedied by using forced C-points.

Nevertheless, also in the case of early Newton steps forced C-points can be used and do not harm the convergence significantly. Hence, if a single approach for a full Newton run shall be used we will suggest PAMG(NO_3^-,max,NO_3^-)-FC(1) for the migration-diffusion system considered. Another possibility is to introduce a control mechanism which monitors the convergence behavior of the PAMG approach. At the beginning PAMG(NO_3^-,max,NO_3^-) is used. When the convergence slows down or even divergence occurs, forced C-points are introduced.

The choice of the partitioning has been made with respect to the geometry, mesh and potential difference between the electrodes. It has to be adapted considering other geometries, meshes, ion-systems and potential differences. However, the latter has only a minor influence, since the migration coefficient increases exponentially towards the electrode. This is also the reason why we are able to choose the same partitioning for two scenarios considered, A and B, despite of their different potential differences.

Chapter 7

A Physics-Aware PAMG Approach for the MITReM

The MITReM describes electrochemical processes via a non-linear PDE system which consists of convection, diffusion, migration and reaction terms. It is discretized with a combined residual-distribution finite element method, and linearized with a Newton method, see Chapter 2.

The standard linear solver used within the MITReM simulator employed is a ILU-GMRES method. Such type of methods are well known to show a super-linearly increasing run time with increasing system size. Hence, especially for industrial relevant applications, the method leads to very long run times of the simulation software. Therefore, we suggest to replace this solver by an AMG method. AMG methods are well known to exhibit optimal complexity and, hence, lower run times for many problem classes, cf. Chapter 3.

The properties of the MITReM and the discretization used lead to convergence problems when applying state-of-the-art PAMG approaches. The problems are caused by the dominance of the convection, the nonlinearities, and the local violation of the Péclet condition.

Considering convection-dominated flow problems, it is well-known that an ordering of the variables in flow direction is beneficial if employing a Gauss-Seidel type smoother for AMG, cf. Chapter 3. In order to find a suited permutation of the variables in the case of the MITReM, we have introduced a **reordering framework** in Chapter 4. The framework makes it possible to define an ordering of variables which is based on the linear system itself and additional external information such as vector fields. Such orderings have been shown to lead to a better representation of the flow than the standard techniques for the MITReM, cf. Section 4.7.

The mesh Péclet number describes the ratio between convection, diffusion and mesh size. Hence, grid information is needed to compute it. However, on coarse levels of the AMG method, no grid information is available besides the couplings between variables. Hence, we have developed a **heuristic Péclet number** for the case of scalar applications and systems of PDEs which makes it possible to compute the Péclet number purely based on algebraic criteria, see Chapter 5. The heuristic Péclet number allows the localization of numerical critical areas on every level of the AMG approach. We have introduced a coarse-grid correction technique which makes use of the Péclet number information and, hence, leads to AMG approaches which are hardly affected by a locally violated Péclet condition.

In addition to the problems caused by the locally violated Péclet condition, convergence problems caused by the non-linearity of the migration-diffusion system have been examined in Chapter 6. The problems are caused by an underrepresentation of the migration especially near the electrodes in the state-of-the-art primary matrix for linear systems from early Newton steps. This can be overcome by using a primary matrix which is adapted to the model problem, i.e. takes the position of the electrodes into account.

Summarizing, we have developed several strategies to deal with the arising difficulties with help of the model problems. The strategies use additional external information which is not included in the linear system to enhance the convergence and robustness of the PAMG approach. Based on the findings for the model problems, we develop robust PAMG variants for the MITReM and investigate their efficiency on a range of industrially relevant geometries and ion-systems in this chapter.

We start with the description of the model problems considered for the MITReM and the employed Newton linearization in Section 7.1. Then, we compare state-of-the-art PAMG approaches with the ILU-GMRES solver of the simulation software. We demonstrate that these PAMG approaches lead to insufficient convergence results in Section 7.2. Hence, we develop a robust and efficient **physics-aware PAMG** approach for the MITReM in Section 7.3. We benchmark the new approach against state-of-the-art PAMG approaches in Section 7.4. Finally, we give important conclusions on the results in Section 7.5.

7.1   The MITReM System

We briefly describe the ion-systems (Section 7.1.1) and geometries (Section 7.1.2) considered and the Newton method employed (Section 7.1.3). Further details regarding the MITReM can be found in Chapter 2.

7.1.1   Ion Systems

ion-system no.	homog. reactions	ions	no. of phys. unknowns
1	-	Ag^+, NO_3^-	3
2	-	$Ag^+, NO_3^-, K+$	4
3	yes	$NaS_2O_3^-, Na^+, S_2O_3^{2-},$ $NO_3^-, AgS_2O_3^-, Ag(S_2O_3)_2^{3-}$	7

Table 7.1: Description of the ion systems.

The ions included in the ion-systems considered are shown in Table 7.1. Note that ion-systems 1 and 2 do not make use of homogeneous reactions, which results in the absence of the reactive term R_i (see Section 2.3, Equation (2.22)). The homogeneous reaction of ion-system 3 is

$$S_2O_3^{2-} + AgS_2O_3^- \rightleftharpoons Ag(S_2O_3)_2^{3-}. \tag{7.1}$$

Ion-system 3 is a realistic silver-ion model [109], the other systems are standard model cases for electrochemical simulators designed to reveal several difficulties, not restricted to, but also for linear solvers.

ion	z_i	D_i
Ag^+	1	1.65E-09
NO_3^-	-1	1.90E-09
K^+	1	1.96E-09
$NaS_2O_3^-$	-1	0.60E-09
Na^+	1	1.33E-09
$S_2O_3^{2-}$	-2	1.90E-09
$AgS_2O_3^-$	-1	0.60E-09
$Ag(S_2O_3)_2^{3-}$	-3	0.60E-09
T		293 Kelvin

Table 7.2: Properties of the ions.

Table 7.2 shows the parameters used for the ions considered. Note that we use the Nernst-Einstein relationship $u_i = D_i/\mathfrak{R}T$ despite the fact that we have non-zero concentrations, cf. Section 2.4.

All models were simulated at 80 percent of the limiting current. The limiting current is approached as the rate of the charge-transfer process is increased by varying the potential. It is independent of the applied potential over a finite range, and is usually evaluated by subtracting the appropriate residual current from the measured total current.

Notation 7.1. *In the following, c(X) denotes the concentration of X, where X is one of the ions included in the ion-system considered.*

7.1.2 Geometries

name	grid points
CHANNEL-SMALL	30,123
CHANNEL-BIG	49,693
BFS-SMALL	31,868
BFS-BIG	54,624
CREVICE-SMALL	36,611
CREVICE-BIG	54,011

Table 7.3: Geometries and grid sizes.

The MITReM is examined on 3 different geometries with two grid sizes each.

- A simple channel geometry, which we investigate to gain insight in the basic solver behavior.

- A backward facing step (BFS) geometry, which is more difficult to solve because large recirculating areas evolve.

- A crevice geometry, which is particularly interesting for investigating corrosion processes.

The geometries and associated grid sizes are listed in Table 7.3. The size of the large grids are (more than) fine enough to resolve all important characteristics of the electrochemical systems considered here.

Channel

Although the channel, see Figure 7.1, is the easiest geometry, a lot of insight can be gained by investigating the solver behavior for this case. The convection and migration contributions are varying slightly in the geometry and are easy to manipulate by adapting the inlet velocity and the electrode potential. Hence, also the dominance of these phenomena can be varied very easily.

In industrial applications, comparable reactors and geometries are used, for example, in plating applications when the electrolyte is refreshed by the forced convection and has a high refresh rate. Hence, the channel test case is even of practical importance.

The channel geometry considered has two electrodes positioned opposite to each other in the middle of the upper and lower boundary of the channel. The channel length is 0.3, its height is 0.03, the upper electrode's length is 0.1, and the lower electrode has a length of 0.08.

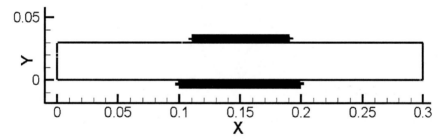

Figure 7.1: Channel geometry with electrodes (gray).

Backward facing step (BFS)

A common geometry considered in fluid dynamics is the backward facing step, see Figure 7.2. The electrodes in this geometry are localized at the reattachment point of the fluid flow. The difficulty with this geometry is the recirculation region with cyclic convection.

The backward facing step (BFS) geometry considered has two electrodes positioned opposite to each other in the middle of the upper and lower boundary. Its length is 0.3, the inlet width is 0.01, the outlet width 0.03, and the step is positioned at 0.05. The upper electrode's length is 0.1, the lower electrode has a length of 0.08. Both are positioned after the reattachment point of the recirculating region.

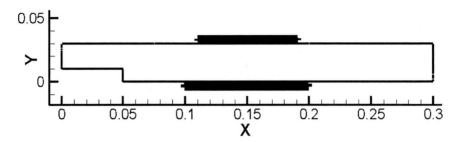

Figure 7.2: Backward facing step geometry with electrodes (gray).

Crevice

The crevice geometry, see Figure 7.3, is particularly used for simulations of electrochemical corrosion processes in materials where, for example, mechanical fatigue is occurring. Inside the crack, a different chemical composition of the electrolyte can occur, which can locally introduce corrosion.

The challenge in performing an electrochemical calculations of a crevice geometry is that the convection inside the crevice is very small or even negligible whereas migration can play an important role in this region, outside the crevice the opposite is true.

The crevice geometry considered has six electrodes. The crack is at the bottom after 0.145 length units with a width of 0.01 and a height of 0.1. One electrode is positioned opposite to the crack at the upper boundary between 0.1 and 0.2. The other electrodes are at the bottom of the geometry and oppositely loaded, one is between 0.11 and 0.145, one between 0.155 and 0.19, and the other three enclose the crack. The hight at the inlet and outlet is 0.03.

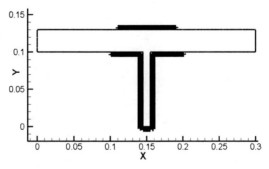

Figure 7.3: Crevice geometry with electrodes (gray).

7.1.3   The Newton Method

The Newton method employed uses a static damping. Hence, it reacts very sensitive to the final residuals of the liner solver. When the linear solver is not able to reduce the residual to a certain threshold, the Newton method is likely to diverge. The initial damping factor is zero, after each nonlinear iteration the factor is increased by 10 percent until the upper bound of 0.8 is reached. We use a relative residual reduction of 6 digits for the full runs unless stated otherwise.

geometry	CHANNEL		CREVICE		BFS	
ion-system	SMALL	BIG	SMALL	BIG	SMALL	BIG
1	38	38	50	50	33	33
2	17	16	19	19	17	17
3	17	17	13	13	14	14

Table 7.4: Number of Newton steps.

Table 7.4 shows the number of non-linear iterations of the Newton method for the ion-systems and geometries considered. The number of non-linear iterations is usually independent of the solver employed. We observe that model problem 1 needs the most Newton steps. This is caused by the very bad initial guess of the Newton method, i.e., the start residual in the first Newton step is 10^4 for ion-system 1 in contrast to 10^{-2} for the other systems.

7.2   State-of-the-Art PAMG for the MITReM

We investigate the performance of state-of-the-art PAMG for the MITReM and compare it to the standard solver employed in the simulation software. The experiments are performed for the simple channel geometry using the small grid size and the realistic crevice geometry using the large grid size.

First, we introduce the solution approaches employed in Section 7.2.1. Then, we compare the performance of the approaches in Section 7.2.2.

7.2.1   Linear Solver

We investigate two state-of-the-art PAMG approaches, which only differ in the choice of the primary matrix and compare them to the standard solver used in the simulation software. One PAMG approach makes use of a norm based primary matrix. This approach will be denoted with PAMGn. The other state-of-the-art approach uses a specified unknown coupling which represents the primary matrix, we denote this approach by PAMGp. The standard solver used in the simulation software is denoted with ILU.

ILU

The standard solver of the simulation software is an ILUT preconditioned GMRES method. The Krylov dimension of the method is 30. The fill-in used is 10 and the drop tolerance is 1e-2 , cf. [91] for more details on ILUT preconditioned Krylov methods.

PAMGn

The approach PAMGn uses a straightforward primary matrix which turns out to be very fast, however, it lacks on stability in the case of MITReM. It is based on the maximum-norm. Namely, the maximal entry of each point-coupling matrix defines the respective entry of the primary matrix. To be more specific, PAMG bases the entries p_{kl} of the primary matrix on norms of the point-coupling matrices $\mathbf{A}_{(k,l)}$

$$p_{kl} = -\|\mathbf{A}_{(k,l)}\| \ (k \neq l) \text{ and } p_{kk} = -\sum_{l \neq k} p_{kl}, \qquad (7.2)$$

where $\|\cdot\|$ denotes the maximum-norm. This choice of the primary matrix is very cheap in terms of computational time.

PAMGp

PAMGp uses a primary unknown. This is also a common technique in the context of PAMG. The unknown-coupling of an ion-concentration unknown to itself defines its primary matrix. PAMGp uses an ion-concentration unknown based primary matrix. We use the concentration coupling of Ag^+ for systems 1 and 2 and the concentration coupling of $NaS_2O_3^-$ for system 3.

Note that the performance of the method depends on the choice of the ion. Especially, choosing ions which take part in the homogeneous reactions will lead to very different results. However, the results do not change significantly when using different unknowns which do not take part in the homogeneous reactions.

7.2.2   Performance Results

geometry	ion-system	PAMGn	PAMGp	ILU
CHANNEL	1	176	295	2005
SMALL	2	169	430	2211
	3	324	824	11292
CREVICE	1	div.	716	div.
BIG	2	538	1036	6311
	3	div.	1414	mem

Table 7.5: Run times in seconds. div. = the Newton method diverged. "mem" = solver does not work (considerably more than 4 GB memory necessary) [105].

Analyzing the run times shown in Table 7.5, we observe that PAMGn and PAMGp outperform ILU considerably. Furthermore, ILU shows divergence for ion-system 1 for CREVICE-BIG and it is not able to solve ion-system 3 for this geometry, too, because of its large memory requirements.

Hence, the PAMG methods are candidates for replacing the original solver. PAMGn is faster than PAMGp, however, PAMGn does not converge for industrial relevant ion-systems and geometries. PAMGp shows a very robust behavior. However, its runtime is often 2 times slower than the run time of PAMGn.

The reason for the insufficient solving properties of the PAMG approaches are the locally violated Péclet condition and the nonlinearity of the system. In the following, we develop physics-aware PAMG approaches which are robust *and* efficient for the MITReM and show their capability in further numerical experiments.

7.3   Derivation of Efficient and Robust PAMG Approaches

Applying state-of-the-art PAMG approaches to the MITReM, we observe a level-dependent convergence behavior for simple geometries and ion-systems and even divergence for more complicated geometries and ion-systems. In the previous chapters, we already pointed out the reasons for this behavior with the help of model problems. We have developed several strategies to deal with the arising difficulties - convection-dominance of the system, local violation of the Péclet condition, and non-linearities. The physics-aware AMG strategies developed use a smoothing which can take the direction of dominant flow into account, a primary matrix which is adapted to the model problem considered, and a special treatment of points violating the Péclet condition.

We investigate the performance of state-of-the-art PAMG approaches and compare it to the physics-aware PAMG approaches for the case of the MITReM. In order to facilitate the analysis, we only consider single matrices in this section. The results for full simulation runs are shown in Section 7.4.

Notation 7.2. *We name the linear systems considered IX-MY, where $X \in \{1, 2, 3\}$ denotes the ion-system and $Y \in \mathbb{N}$ the Newton step, e.g., I1-M5 denotes the 5th linear system from the Newton method using ion-system 1.*

In this section, we use the term "convergence factor", which denotes the average reduction of the residual for iterations 3 to N, where N is the number of iterations performed. The first two iterations are left out, since they usually show a "random" convergence behavior.

The abbreviation "div." used in the following tables denotes that the respective method diverged.

In Section 7.3.1, we introduce the basic PAMG components which the PAMG approaches considered have in common. A special focus is on the choice of an appropriate accelerator, aggressive coarsening strategies, and the direct solver on the coarsest level. In Section 7.3.2, we investigate the behavior of several smoothers and orderings of variables on the convergence of PAMG in the context of the MITReM, the orderings are determined by means of the reordering framework. Finally, we introduce the physics-aware coarsening and investigate which type of coarsening is suited best for which ion-system in Section 7.3.3.

7.3.1   The Basic PAMG Components

We introduce the basic PAMG components to be used by all PAMG approaches considered. The choice of the basic components is motivated by numerical experiments using a straightforward PAMG approach called PAMGn which we introduce in detail in the following.

Smoother

The PAMG approaches employed make use of block-ILU (BILU) or block-Gauss-Seidel (BGS) smoothing. The choice of a good smoother is a prerequisite for an efficient multigrid method. The smoother should reduce the "high" error frequencies and, in particular, be cheap in terms of memory requirement. Since the MITReM is a strongly coupled PDE system, smoothers like Gauss-Seidel or ILU(0) which are frequently employed in AMG do not work (efficiently). Hence, we make use of their respective block variants, where a block is defined to be a point-coupling matrix, cf. Section 3.2.2.

All PAMG approaches make use of BILU smoothing because the BGS smoother strongly diverges on coarse levels which leads to divergence of the full approach. The reason being the large Péclet numbers. The smoother behavior is discussed in detail in Section 7.3.2.

Coarsening and Interpolation

The coarsening and interpolation employed in the PAMG approaches is based on the primary matrix, see Section 3.2 for details. Furthermore, the primary matrix used for coarsening is always the same as the one used for interpolation. The interpolation is of single-unknown type.

The straightforward PAMG approach bases the primary matrix on the maximum-norm. Using the maximum-norm is a computationally cheap and commonly used way to compute the primary matrix. It means that the maximal entry of each point-coupling matrix defines the respective entry of the primary matrix. To be more specific, PAMG bases the entries p_{kl} of the primary matrix on norms of the point-coupling matrices $\mathbf{A}_{(k,l)}$

$$p_{kl} = -\|\mathbf{A}_{(k,l)} \ (k \neq l)\| \text{ and } p_{kk} = -\sum_{l \neq k} p_{kl}, \qquad (7.3)$$

where $\|\cdot\|$ denotes the maximum-norm.

Coarsest Level Solver

We compare several common coarsest level solver for the MITReM using the straightforward approach PAMGn. The aim is to find the optimal coarsest level solver with respect to run time, memory requirements and robustness. The solvers considered are the direct solvers PARDISO [53, 92] and MUMPS [1], and the one-level iterative solvers BICGSTAB-ILUT(a,b)

geometry	matrix	solver			
	of level	PARDISO	MUMPS	GMRES	BICGSTAB
CHANNEL	2	2.2E-08	1.6E-08	1.0E-10	8.1E-11
	3	1.1E-08	2.2E-08	6.0E-12	7.7E-12
	6	2.0E-13	2.2E-09	3.5E-07	5.1E-14
BFS	2	3.8E-08	9.1E-10	2.3E-09	1.1E-09
	3	7.9E-11	1.1E-08	5.7E-11	5.1E-11
	6	6.6E-12	1.0E-08	9.1E-13	3.6E-12

Table 7.6: Output residuals for system I1-M5.

(BICGSTAB) and GMRES-ILUT(a,b) (GMRES). BICGSTAB-ILUT(a,b), denotes an ILUT preconditioned BICGSTAB method where a is the absolute fill-in and b the threshold value, see [91] for details on the iterative solvers.

The Tables 7.6 and 7.7 show output residuals for $a = 50$ and $b = 0.0$ for the backward facing step as well as the channel geometry. The Krylov dimension for the GMRES solver is set to 50. We use a random first guess vector such that the starting residual is one, and a right hand side vector of ones. The iterative methods are stopped if the machine accuracy or 150 iterations are reached.

In Table 7.6, all coarse level solver show a sufficient reduction of the residual. Note that for most applications a reduction of two or three orders of magnitude on the coarsest level is sufficient. However, when dealing with points/variables which violate the Péclet condition this is not the case, especially, since the smoother strongly diverges for these points/variables.

The table also shows an improvement in the residual reduction for coarse level matrices between the second and third level for all solvers. However, this improvement only shows up for ion systems 1 and 2. It might be caused by the different coupling structures of the coarse level matrices. Namely, the share of the concentration-to-concentration couplings in the maximum-norm based primary matrix increases between levels 2 and 3 for ion systems 1 and 2.

Considering Table 7.7 which shows output residuals for ion-system 3, we observe a lower accuracy of the solutions returned by the solvers considered compared to ion-systems 1 and 2 for the matrices of the second and third level. This is especially true for the BFS geometry, see Table 7.7. The Krylov methods report a numerically singular input matrix for the third level matrix of the BFS geometry. This can only be avoided if the fill-in

geometry	matrix	solver			
	of level	PARDISO	MUMPS	GMRES	BICGSTAB
CHANNEL	2	5.6E-08	7.7E-08	3.7E-05*	6.4E-03*
	3	2.1E-08	3.4E-08	1.4E-04*	2.1E-02*
	6	4.3E-13	1.3E-09	6.1E-11	1.4E-05*
BFS	2	5.1E-03	4.1E-06	1.9E-03*	9.3E+06*
	3	2.4E+02	7.6E-06	-	-
	6	1.1E-08	1.5E-07	1.1E-00*	9.9E-01*

Table 7.7: Output residuals for system I3-M5; (*): all 150 iterations performed; (-): preconditioned matrix is numerically singular.

is increased. Namely, two times the average number of entries per level is needed to achieve convergence in this case.

Table 7.7 also shows that PARDISO cannot solve the 3rd coarse level matrix of BFS for ion-system 3. The reason for this is rooted in PARDISO's pivoting strategy which is only one-sided where MUMPS uses a multi-frontal approach.

Table 7.8 compares the memory and run time results of PARDISO and MUMPS. MUMPS shows a good overall behavior, however, it has a rather large memory requirement for small matrices. Overall, the direct solver MUMPS provides a good compromise between numerical robustness and run time. However, since PARDISO is mostly faster and needs less memory than MUMPS, especially, for small matrices, we mainly use PARDISO as a coarsest level solver. When PARDISO is not able to solve the coarsest linear system sufficiently well we automatically switch to MUMPS. This strategy is used in all PAMG approaches considered for the MITReM.

matrix	# elements	PARDISO		MUMPS	
of level	per row	mem.	T	mem.	T
2	43	806	12	837	30
3	60	251	9	235	8
6	47	3	0	39	0

Table 7.8: Run times in sec. (T) and memory requirements in MB (mem.) for system I3-M5 using BFS-BIG.

Acceleration

AMG preconditioned Krylov methods, like CG (only for symmetric matrices), BICGSTAB, or GMRES often show a much better robustness and convergence than stand-alone AMG approaches. From the AMG point-of view, Krylov methods enhance AMG's convergence. Hence, they are often called accelerators in the context of multigrid methods.

Furthermore, in many industrial applications which usually do not fulfill the propositions of algebraic multigrid methods, the use of an accelerator for AMG is a prerequisite for an efficient and robust solution approach. Hence, we investigate the effect of an accelerator for the MITReM.

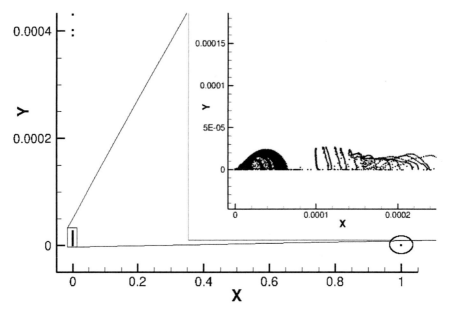

Figure 7.4: Typical eigenvalue distribution of an iteration matrix for the discretized electrochemical system I1-M5.

Figure 7.4 shows a representative spectrum of the eigenvalues of system I1-M5 for a very small channel geometry with just 1000 points. We observe that there are some eigenvalues without imaginary part which are large (marked with the ellipse). The other eigenvalues are clustered in the positive hemisphere around zero. Krylov methods are known to handle such large real eigenvalues efficiently. Hence, the use of Krylov methods usually enhances the convergence of the PAMG approaches.

Table 7.9 compares convergence factors and memory requirements of

	no accel.		GMRES		BICGSTAB	
system	conv.	MB	conv.	MB	conv.	MB
I1-M5	0.48	153	0.24	173	0.10	162
I1-M30	0.71	158	0.30	173	0.17	162
I3-M2	0.95	464	0.59	499	0.22	475
I3-M10	0.67	464	0.30	499	0.13	475

Table 7.9: PAMGn results using different accelerators. conv.: convergence factor. MB: peak memory in MB.

the non-accelerated PAMGn method, with the GMRES-accelerated and the BICGSTAB-accelerated one. The convergence factors improve drastically when using an accelerator as we would expect from the eigenvalue spectrum.

Note that the convergence of BICGSTAB is better than in the case of GMRES. However, BICGSTAB performs two inner AMG iterations per BICGSTAB iteration where GMRES only performs one. Hence, the run-time per BICGSTAB iteration is larger than the run-time per GMRES iteration. Overall, both methods achieve very similar run-times. We use BICGSTAB in all PAMG approaches considered rather than GMRES because of BICGSTAB's lower memory requirements.

Aggressive Coarsening

For some applications, it is beneficial to change the standard coarsening to an aggressive one in order to enhance the runtime, memory requirements, and robustness of AMG. This is especially true when considering small stencil sizes. We investigate whether aggressive coarsening also helps to reduce memory requirements and to maintain a good convergence in the case of the MITReM.

	BILU		PAMGn-agr		PAMGn-std	
system	conv.	MB	conv.	MB	conv.	MB
I1-M5	0.89	66	div.	102	0.10	162
I1-M30	0.92	66	0.83	102	0.17	162
I3-M2	0.95	209	0.91	292	0.22	475
I3-M10	0.97	209	0.92	293	0.13	475

Table 7.10: Comparison of PAMGn-BICGSTAB using standard and aggressive coarsening, and BILU-BICGSTAB. conv. : convergence factor. MB: peak memory in MB.

We employ A1-aggressive coarsening on the first level, the subsequent levels use standard coarsening, see [100] for details on A1-coarsening. The memory requirements shown in Table 7.10 are typical for the methods employed. Namely, PAMG's memory requirement is slightly more than twice as large as in the case of a one-level iterative solver using standard coarsening. Furthermore, PAMG's memory requirements can be lowered drastically when using aggressive coarsening.

Table 7.10 also shows the convergence factors for the BILU-BICGSTAB one-level iterative solver for comparison. The convergence factors of BILU are far worse than the ones of the multigrid method using standard coarsening. This is usually expected for one-level iterative solvers, and also demonstrates the complexity of the problem to be solved.

The table shows that the gain in aspects of memory for the aggressive variant does not pay-off since the method does not converge or converges very slowly compared to the standard coarsening variant. The reason being, the locally violated Péclet condition for the problems considered which increases drastically on coarse levels if employing aggressive coarsening. Consequently, we do not use aggressive coarsening variants for the PAMG approaches considered because of their insufficient robustness and convergence behavior.

7.3.2   The Effect of Reordering

Considering convection-dominant systems, the smoother has to capture the direction of convection appropriately. Ensuring this is still a nontrivial task in practice, and particularly in the case of the MITReM not that obvious. A reason being that, in addition to the convective term, we have to deal with migration which dominates in certain parts of the domain depending on the concrete electrochemical system (ions etc.) and the geometry.

Migration has a convective and diffusive character. However, the movement caused by migration is usually directed differently to the convection of the MITReM. Hence, an "optimal" smoother shall handle all these at least partly conflicting terms appropriately. Furthermore, for many geometries of practical relevance like the ones presented here, the electromagnetic field can even be nearly orthogonal to the direction of the fluid flow (see Figure 7.5). Additionally, the potential gradient which describes the electromagnetic field develops during the simulation process. Therefore, the numerical behavior of the matrices changes within the Newton process because of the changing potential gradient. In particular, this makes it difficult

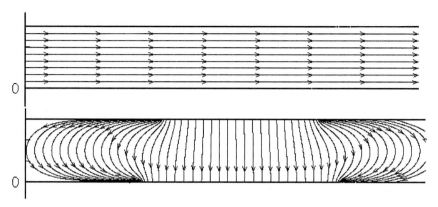

Figure 7.5: CHANNEL-SMALL. Velocity field (top), and electromagnetic field (bottom).

to correctly detect the electromagnetic field during the first Newton steps.

However, for most MITReM simulations, convection is far more dominant in the majority of the computational domain since a uniform distribution of the ions in the solution usually has to be ensured. Furthermore, the migration is even lower than the diffusion in most parts of the domain. Hence, we will get a sufficiently robust behavior of smoothers if we consider the direction of overall flow. That is, we consider a mixture of convection and migration to determine a flow-based ordering of points.

We investigate the effect of three different orderings of variables on the smoothing properties of the BGS and BILU smoother and the interplay between coarse-grid correction and smoothing, namely, the default reverse Cuthill-McKee ordering of the simulator, a state-of-the-art flow-based ordering, and an ordering of variables in the direction of dominant-flow derived by using vector fields. The latter two orderings are computed with the reordering framework proposed in Chapter 4.

Remark 7.3. *We consider the linear system from the 5th Newton step of ion-system 1 (I1-M5) using the CHANNEL-SMALL geometry. In the case of the channel geometry, convection is far more dominant than migration all over the domain except for the part of the domain near the electrode boundaries.*

The Flow-Based Orderings

In Section 4.6, we have demonstrated that an ordering of variables in the direction of dominant flow is beneficial for the convergence of AMG when using a Gauss-Seidel smoother. However, the ordering of the points produced by the grid-generator of the MITReM simulator is computed by means of a reverse Cuthill-McKee algorithm which does not usually represent the direction of dominant flow.

We investigate two flow-based orderings. One ordering is computed with the reordering framework using state-of-the-art techniques (FLOW1), and one ordering additionally uses the new physics-based reduction techniques (FLOW2). The ordering of points using state-of-the-art techniques is computed using a Frobenius-norm basic matrix, cf. Chapter 4. Then, the weighted block-triangular sorting algorithm is applied to compute the final ordering.

We describe the computation of the other permutation which uses the physic-based reduction technique. In previous chapter, we already pointed out that the migration term can be split in a convective and diffusive part. The convective part of the migration is included in the couplings of an ion-concentration to itself and the diffusive term in the potential to ion-concentration coupling, similar to the case of the migration-diffusion system, see Section 6.1.

Since the convection is also included in the coupling of an ion-concentration to itself, both directions of global flow are included in theses couplings. Hence, it is possible to determine the direction of dominant flow solely by a basic matrix governed by an ion-concentration basic unknown.

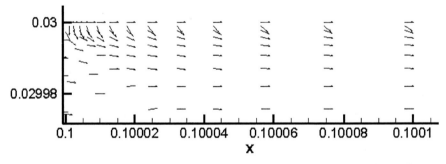

Figure 7.6: Dominant direction of movement for $c(NO_3^-)$ to $c(NO_3^-)$ coupling (detail).

The direction of dominant flow using $c(NO_3^-)$ as basic unknown is visualized in Figure 7.6. Note that the direction of movement caused by the migration for cations, i.e. positively loaded ions, is in the opposite direction.

The ordering of variables in the direction of dominant flow is derived using the $c(NO_3^-)$ unknown as a basic unknown for the basic matrix of the reordering framework. Then, the basic matrix is used to compute the vector field which describes the flow with help of the AV method. Finally, based on the vector field, a reduced basic matrix is computed. Afterwards, the permutation is computed applying the triangularization algorithm to the reduced basic matrix, see Chapter 4 for more details.

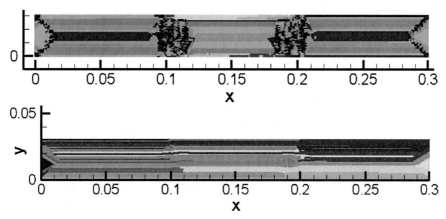

Figure 7.7:   Flow-based ordering of variables using FLOW1 (top) and FLOW2 (bottom). Ordering proceeds from blue to red.

Both flow-based orderings are visualized in Figure 7.7. The ordering of points is visualized by colors. The point with the lowest index is colored dark blue, and the point with the largest index is colored red. Because of the large number of points, the figure only visualizes the quality of the ordering, and not the direction. We use these orderings and the original RCM ordering and compare their effect on the PAMG smoothers BGS and block-ILU (BILU).

Smoothing Behavior of One-Level Iterative Solvers

Considering the smoothing behavior for the concentration couplings $(c(Ag^+)$ chosen exemplary) shown in Figure 7.8, the error reduction proceeds in the direction of convection for BILU using the RCM ordering. This is also observed when using a flow-based ordering, however, in this case the error reduction is not that good than in the case of the RCM ordering. This is expected from the general observations using ILU-type iterative methods with RCM ordering because this ordering reduced the bandwidth of the linear system, cf. Section 4.2.

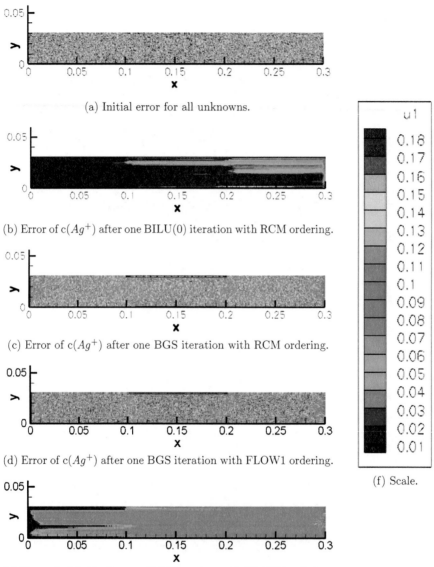

(a) Initial error for all unknowns.

(b) Error of c(Ag^+) after one BILU(0) iteration with RCM ordering.

(c) Error of c(Ag^+) after one BGS iteration with RCM ordering.

(d) Error of c(Ag^+) after one BGS iteration with FLOW1 ordering.

(e) Error of c(Ag^+) after one BGS iteration with FLOW2 ordering.

(f) Scale.

Figure 7.8: Smoothing behavior of the BILU(0) and BGS smoother for the c(Ag^+) unknown.

In case of the BGS smoother, we hardly see any smoothing behavior using RCM ordered variables. The same is true for the FLOW1 ordering which uses state-of-the-art reordering techniques. Considering the FLOW2 ordering, however, the BGS smoothing proceeds in the direction of dominant flow similar to the case of BILU. Note that we use only full smoothing steps in the analysis, because C/F-smoothing would destroy the computed ordering and, hence, prevent good smoothing properties.

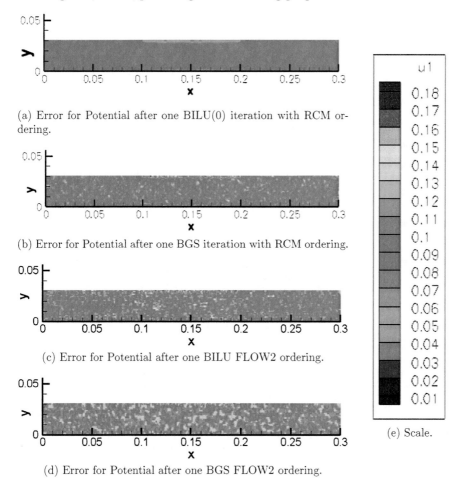

(a) Error for Potential after one BILU(0) iteration with RCM ordering.

(b) Error for Potential after one BGS iteration with RCM ordering.

(c) Error for Potential after one BILU FLOW2 ordering.

(d) Error for Potential after one BGS FLOW2 ordering.

(e) Scale.

Figure 7.9: Smoothing behavior of the BILU and BGS smoother for the potential unknown.

Figure 7.9 shows the impact of the smoothers and ordering of variables on the potential unknown. The equation for the potential unknown is a Poisson equation, see Section 2.3. Hence, the smoothers have better smoothing properties than for the equations of the concentrations, which include the convective, reactive, migrative, and diffusive terms. Therefore, we observe a much better smoothing behavior for all orderings and smoothers considered than in the case of the ion-concentrations. In terms of error reduction, however, the RCM ordering for the BILU smoother is again the best as we have already seen for the concentration equations. Note that we do not show further smoothing images of FLOW1 because of its insufficient smoothing behavior for the ion-concentrations.

We have investigated the smoothing behavior of the BGS and BILU smoother. The BGS smoother shows good performance if the ordering of variables is in the direction of dominant flow and the physics based reduction is used. Using state-of-the-art flow-based reordering or the initial RCM ordering in conjunction with BGS smoothing has no smoothing effect on the error for the concentration couplings. Considering BILU smoother, the smoothing behavior is nearly independent of the ordering considered. Additionally, the error reduction is much better in the case of a RCM ordering.

Interplay Smoothing - Coarse-Grid Correction

smoother	ordering	sweep-type	conv. factor
BGS	RCM	C/F	div.
BGS	FLOW1	C/F	div.
BGS	FLOW2	C/F	div.
BGS	RCM	Full	div.
BGS	FLOW1	Full	div.
BGS	FLOW2	Full	0.32
BILU	RCM	Full	0.13
BILU	FLOW	Full	0.73

Table 7.11: Convergence factors for the PAMGn-2.

We further analyze the dependency of the smoothers on the ordering of variables using the straightforward PAMG two-level method, PAMGn-2. Table 7.11 shows the convergence factors of PAMGn-2 using various smoothing setups. C/F denotes that first the coarse level variables and then the fine level variables are smoothed, also known as red-black smoothing. This smoothing type is usually used within the context of AMG and Gauss-Seidel type smoothing. "Full" denotes a full iteration of the smoother considered.

According to the analysis of the smoothing properties on the error, the best convergence factor is achieved using BILU smoothing with the initial RCM ordering of variables. The second best smoothing method is BGS smoothing using the FLOW2 ordering and the full Gauss-Seidel sweep. PAMGn-2 diverges using the C/F sweep because the splitting of variables partially destroys the flow-wise ordering of variables. Furthermore, we observe that the BGS smoother using the state-of-the-art FLOW1 ordering does not converge, which is a result of its insufficient smoothing properties for the ion-concentrations as we already observed above. Using the RCM ordering in conjunction with BGS smoothing diverges, too, independent of the sweep type. The third best method is employing the BILU smoothing with flow-wise ordering. We have already seen that the error reduction of BILU is far worse using a flow-wise ordering than in the case of the RCM ordering of variables.

We have demonstrated that the BGS smoother requires a flow-wise ordering of the variables which sufficiently well represents the flow direction and that it has to perform full smoothing steps in order to give sufficient results. Such orderings of variables can only be computed for the MITReM using the physics-based reduction techniques introduced.

The behavior of the BGS smoother leads to the assumption that difficulties on the second level may occur using the ordering induced by the fine level. Additionally, either the convection or the migration may become more dominant on the second level which might lead to a different character of the direction of dominant flow.

The reordering framework introduced in Chapter 4 is able to create a flow-ordering, separately, for each level. We investigate the behavior of the smoothers for the induced ordering from the previous level, and a newly computed ordering on the second level. For the BILU smoother we work with an RCM ordering of variables, for the BGS smoother we use the FLOW2 ordering.

Figure 7.10 shows the smoothing behavior of BILU and BGS for the linear system of the second level. Two different orderings are used, the one induced by the first level (either RCM or FLOW2), and a newly computed one solely based on the second level information. In the case of BILU, the newly computed RCM ordering is determined with help of a basic matrix which is governed by the maximum-norm of the point-coupling matrices. In the case of the BGS smoother, the newly computed flow-wise ordering is determined using the same components of the reordering framework as in the case of FLOW2.

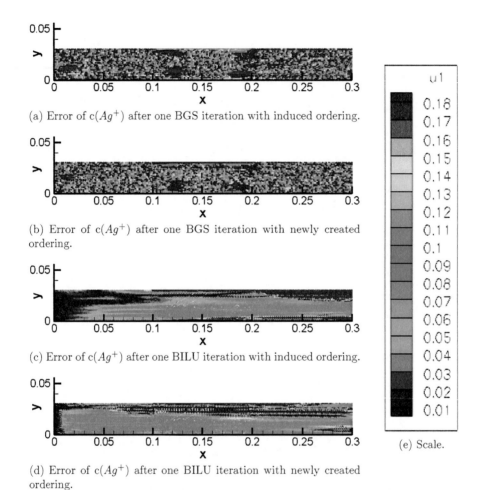

(a) Error of c(Ag^+) after one BGS iteration with induced ordering.

(b) Error of c(Ag^+) after one BGS iteration with newly created ordering.

(c) Error of c(Ag^+) after one BILU iteration with induced ordering.

(d) Error of c(Ag^+) after one BILU iteration with newly created ordering.

(e) Scale.

Figure 7.10: Smoothing behavior of the BILU(0) and BGS smoother for the c(Ag^+) unknown on level 2.

We observe that the BGS smoother does not smoothen the error visibly for both orderings and, furthermore, strongly diverges. This is caused by the distribution of large Péclet numbers on the second level, we give more details on the distribution of large Péclet numbers on coarse levels in Section 7.3.3. In contrast to BGS, BILU shows a better smoothing behavior for both orderings. However, the error reduction seems to work better in the case of the induced RCM ordering from the fine level.

smoother	ordering	conv. factor
BGS	FLOW2-induced	div.
BGS	FLOW2-newly computed	div.
BILU	RCM-induced	0.19
BILU	RCM-newly computed	0.73

Table 7.12: Convergence factors for PAMGn-3.

Considering the convergence factors for the three-level method, PAMGn-3, shown in Table 7.12, we see that the insufficient smoothing behavior of the BGS smoother leads to a diverging approach. Furthermore, the better error reduction of the BILU smoother, in case of the induced ordering, is displayed well in the convergence factor.

Note that the smoothers should smooth the error, and do not necessarily have to converge. However, if a linear system contains global flow it is often advantageous to have a smoother which solves the "convective part" of the error. This is the reason why the good error reduction of the smoother also displays well in the overall convergence factor.

Conclusion

In the following, we only consider BILU smoothing because the BGS smoother does not work using more than 2 levels. Furthermore, we have demonstrated that the BILU smoother does not benefit from a reordering of variables on coarse levels. Hence, we always use the induced ordering from the previous level. We already pointed out that the grid-generator of the simulation software employed produces a RCM ordering. However, often the starting point of this ordering is not at the inlet of the domain. In this case, we reorder the variables in a RCM manner using a starting point which is positioned at the inlet.

The convergence factor of the three-level BILU-PAMGn method using the induced ordering is slightly worse than in the case of 2 levels. Table 7.13

approach	levels					
	1	2	3	4	5	6
BILU-PAMGn	0.96	0.13	0.19	0.23	0.31	0.48

Table 7.13: Convergence factors for system I1-M5.

shows that the convergence factor even worsens if we apply more levels. This level-dependency is caused by large Péclet numbers on coarse levels, which also occurred in the case of the migration-diffusion systems, cf. Chapter 6. In the next section, we further investigate this issue, and introduce a physics-aware coarsening which takes large Péclet numbers into account in order to prevent a level-dependency of the approaches.

7.3.3   Introduction of a Physics-Aware Coarsening

In the case of the migration-diffusion system, we observe that the migration at the electrodes is underrepresented for early Newton steps where the approximation of the potential unknown is far away from the solution when considering a maximum-norm based PAMG approach, see Chapter 6. Nonetheless, the migration has a strong influence to the overall flow at the electrodes. Hence, it should be considered in the PAMG approach employed.

We have demonstrated that a physics-aware primary matrix shows a much better convergence behavior than state-of-the-art primary matrices. The physics-aware primary matrix proposed for migration-diffusion systems uses an ion-concentration unknown based primary matrix near the electrodes and a maximum-norm based primary matrix in the remaining part of the domain.

In the case of the MITReM, the migration has a strong effect on the dominant flow near the electrodes, too. We have already pointed this out in the previous section, cf. Figure 7.6. Hence, we investigate if the migration is also underrepresented for early Newton-steps in the case of the MITReM. Exemplary, we analyze the situation for the CHANNEL-SMALL geometry using ion-system 1.

Figure 7.11 shows the dominance of several unknown-couplings in the point coupling matrices in the part of the domain near the upper electrode. A value of 1 means that only this coupling is present, and a value of 0 means that the coupling entry is zero. For example, (b) shows the coupling from the NO_3^--concentration to the potential unknown for system I1-M5. This coupling dominates nearly everywhere except for the electrode boundary.

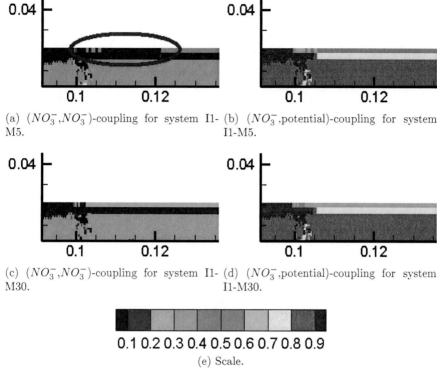

(a) (NO_3^-, NO_3^-)-coupling for system I1-M5.

(b) $(NO_3^-$,potential$)$-coupling for system I1-M5.

(c) (NO_3^-, NO_3^-)-coupling for system I1-M30.

(d) $(NO_3^-$,potential$)$-coupling for system I1-M30.

0.1 0.2 0.3 0.4 0.5 0.6 0.7 0.8 0.9

(e) Scale.

Figure 7.11: Dominance of couplings in maximum-norm based primary matrix at each point of the domain (detail).

Hence, its coupling is represented in the majority of entries in a maximum-norm based primary matrix. This observation is similar to the case of the migration-diffusion equation.

The velocity in the interior of the domain is larger than near the boundary. Consequently, the coupling dominance of the $(c(NO_3^-), c(NO_3^-))$-coupling which includes the convective term shows only dominance in the interior. At the electrodes the $(c(NO_3^-), c(NO_3^-))$-coupling dominates because of the strong migration. However, for system I1-M5 the dominance is not visible near the whole upper electrode, marked with the ellipse. Hence, we expect that the migration will be underrepresented in this area if the primary matrix of system I1-M5 is derived by the maximum-norm. The reason for this effect is the approximation of the potential gradient which is far away from the final solution for the early systems.

Overall, the results indicate that the flow described by the migration in

the direct surrounding of the electrodes is underrepresented in early Newton steps, like system I1-M5. This effect has also been observed for linear
systems from early Newton steps in the case of the migration-diffusion equation, see Chapter 6. We demonstrate that using a primary matrix derived
by an ion-concentration as primary unknown near the electrodes and the
maximum-norm based primary matrix for other points might improve the
convergence and, especially, reduce the level-dependency already observed
in Table 7.13.

The geometry and grid considered is the same as in the case of the
migration-diffusion equation. Hence, we use primary matrices which are
similar to those used in that case, cf. Section 6.3. The methods use a
splitting of the domain into three parts, one part around the upper electrode Ω_1, one around the lower electrode Ω_3, and a third part containing
the remaining grid points Ω_2. Then, the general approach PAMG(A,B,C)
uses the concentration of ion A (B,C) as a primary unknown in Ω_1 (Ω_2,
Ω_3), if A,B, or C are ions. In case of A=max (B=max,C=max) the primary matrix is governed by the maximum-norm. For example, the method
PAMG(NO_3^-,max,NO_3^-) uses the c(NO_3^-) unknown as a primary unknown
near the upper and lower electrode, and the maximum-norm based primary
matrix for the other points. In the case of the channel geometries, the domain around the electrodes have a rectangular shape with height $2r$ and
length $l_e + 2r$, where r is 0.00002 and l_e is the electrode length.

An optimal partitioning of the domain depends on the geometry, mesh,
potential difference between the electrodes, as well as the velocity field.
However, since all simulation setups use a limiting current of 80%, and
the migration coefficient has a steep gradient near the electrodes, we only
consider the position of the electrodes and the direction of dominant flow
in order to find a good partitioning of the domains considered.

We propose several physics-aware primary matrices for the ion-systems
considered and show their benefits for single matrices extracted from the
Newton process using the CHANNEL-SMALL geometry. Furthermore, we
analyze several techniques to deal with a locally violated Péclet condition.
The choices of the partitioning for the other geometries considered are discussed in Section 7.4.

Ion-System 1

We demonstrate the benefits of a physics-aware coarsening for ion-system 1
using the 5th and the 30th linear system from the Newton method for the

channel geometry. In the case of ion-system 1, the Newton method takes 37 iterations to converge. The 5th linear system (I1-M5) is far away from the last system, whereas the 30th system (I1-M30) is qualitatively similar to the last system. This allows us to analyze the effect of the physics-aware coarsening for linear systems from early and late Newton steps, and compare the results to the migration-diffusion system described in Chapter 6.

We have demonstrated that the approach PAMG(NO_3^-,max,NO_3^-) is nearly level-independent considering systems from early Newton steps of the migration-diffusion equation. The approach uses the NO_3^--concentration as a primary unknown near the upper and lower electrode, and the maximum-norm based primary matrix for the other points. We have performed numerous benchmarks with different partitionings and the methods used in the context of the migration-diffusion system, cf. Section 6.3. It has turned out, that the choice of the primary matrix of PAMG(NO_3^-,max,NO_3^-) leads to the best convergence factors in the case of the MITReM, too. This is rooted in the similarities of the systems of PDEs. Hence, we use only this physics-aware primary matrix and compare it to the straightforward approach, here.

approach		levels				
		2	3	4	5	6
I1-M5						
PAMG(max)		0.13	0.19	0.23	0.31	0.48
PAMG(NO_3^- max NO_3^-)		0.18	0.19	0.24	0.33	0.37
I1-M30						
PAMG(max)		0.27	0.31	0.43	0.53	0.71
PAMG(NO_3^- max NO_3^-)		0.43	0.52	0.61	0.65	-

Table 7.14: Convergence factors for systems I1-M5 and I1-M30.

Table 7.14 compares the convergence factors of the straightforward P-AMG(max) method to the PAMG(NO_3^-,max,NO_3^-) method. Apparently, the level-dependency for the system I1-M5 is stronger for the PAMG(max) method than in the case of the PAMG(NO_3^-,max,NO_3^-) method.

Considering the late system, I1-M30, the PAMG(NO_3^-,max,NO_3^-) approach has the advantage that it performs only 5 levels, which shows a better convergence factor than the 6-level PAMG(max) approach. However, its 5-level convergence factor is worse than the 5-level factor of PAMG(max). The reason for the better convergence of the 5-level PAMG(max) method compared to PAMG(NO_3^-,max,NO_3^-) is that the potential unknown is approximated well at this stage of the Newton process for system I1-M30. Hence, the migration is represented sufficiently in the maximum-norm based primary matrix, cf. Figure 7.11. Additionally, PAMG(NO_3^-,max,NO_3^-) completely neglects the diffusion of the ion-concentration to potential coupling near the electrode which causes an disadvantageous coarsening, as we have already observed in the case of migration-diffusion.

Figure 7.12 shows the Péclet numbers for systems I1-M5 and I1-M30 using PAMG(max) on level 2 and 5. The Péclet numbers at the electrode boundary on the fine level are very large in both cases. These large Péclet numbers are caused by an insufficiently refined grid which has many small angles near the electrode because it is highly refined in y-direction and not simultaneously in x-direction.

Furthermore, similar to the migration-diffusion system, the Péclet numbers increase the better the final solution is approximated. Hence, we observe that the Péclet numbers for system I1-M30 are larger than for system I1-M5. Using PAMG(max), the large Péclet numbers spread out on coarse levels in the case of system I1-M30. In contrast to this, the points which violate the Péclet condition vanish for level five of A-5, see Figure 7.12.

The significant worsening of the convergence factor from the 5 to the 6-level approach of PAMG(max), for both systems considered, is caused by the smoother on level 5. The smoother oscillates at the electrodes, see Figure 7.13. The oscillations are caused by the grid, which is too coarse near the electrodes causing very large Péclet numbers, cf. Figure 7.12.

We conclude that a physics-based primary matrix helps to improve the convergence of the straightforward approach. However, the approach maintains its level-dependency.

In the case of the migration-diffusion system, we have shown that the use of forced C-points helps to lower the level-dependency and enhances the convergence, especially, for systems from late Newton steps, see Section 6.4.2. The aim of using forced C-points is to avoid increasing Péclet numbers on coarser levels. The proposed PAMG(*)-FC(*) methods analyze the Péclet numbers on each level and force points which violate the Péclet condition, i.e. critical points, to the coarsest level.

We consider two variants of the PAMG(*)-FC(*) approach. The P-AMG(*)-FC(0) method forces only the critical points to the coarsest level, and PAMG(*)-FC(1), additionally, forces the direct neighbors of all critical points to the coarsest level.

Table 7.15 shows the convergence factors for the linear systems considered using forced C-points. In the case of system I1-M5, the use of forced

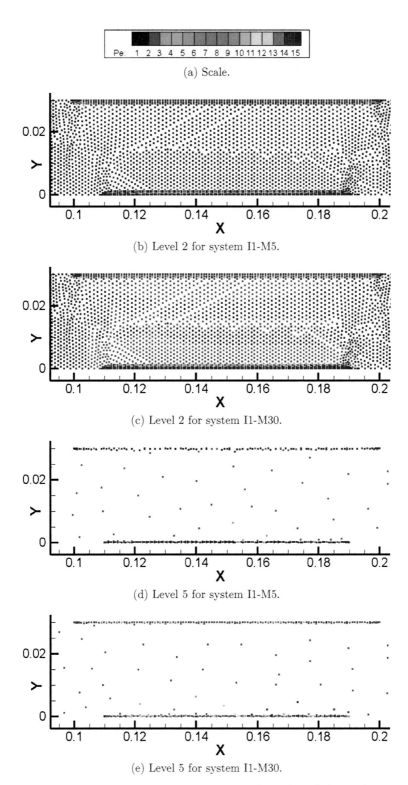

(a) Scale.

(b) Level 2 for system I1-M5.

(c) Level 2 for system I1-M30.

(d) Level 5 for system I1-M5.

(e) Level 5 for system I1-M30.

Figure 7.12: Péclet numbers using PAMG(max) (detail).

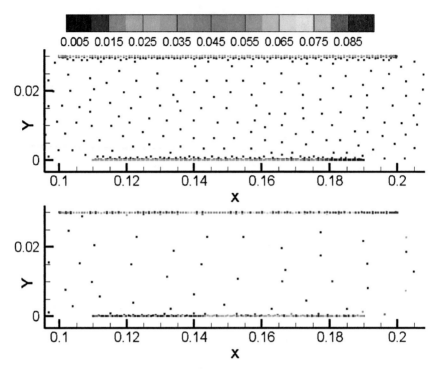

Figure 7.13: System I1-M30: Error after 5 smoothing steps on level 4 (top) and level 5 (bottom), starting with a random first guess.

approach		levels		
		2	3	4
I1-M5				
PAMG(max)-FC(0)		0.12	0.18	0.27
PAMG(NO_3^- max NO_3^-)-FC(0)		0.19	0.20	0.33
PAMG(max)-FC(1)		0.13	0.20	0.24
PAMG(NO_3^- max NO_3^-)-FC(1)		0.19	0.21	0.25
I1-M30				
PAMG(max)-FC(0)		0.13	0.20	0.31
PAMG(NO_3^- max NO_3^-)-FC(0)		0.16	0.23	0.24
PAMG(max)-FC(1)		0.13	0.17	0.21
PAMG(NO_3^- max NO_3^-)-FC(1)		0.13	0.22	0.22

Table 7.15: Convergence factors for systems I1-M5 and I1-M30 using forced C-points.

C-points does not improve the convergence of the 4-level methods. However, the coarsening ends at level 4 because the coarsening is slow. Hence,
the technique helps to identify where the coarsening shall stop in order to
remedy slow convergence.

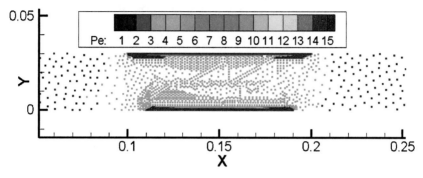

Figure 7.14: System I1-M30: Péclet numbers of level 4 using PAMG(max)-
FC(0) (detail).

For system I1-M30, not only the number of levels performed is lower,
but also the convergence factors for the 4-level approaches improve when
using forced C-points. This is due to the fact that no additional points with
large Péclet numbers arise, see Figure 7.14.

Similar to the results of the migration-diffusion system, we have demonstrated that the use of forced C-points only improves the convergence in the
case of system I1-M30. This is caused by the Péclet numbers on the first
level which are lower for system I1-M5 than for the late one system I1-M30.

Table 7.15 shows that it pays off to, additionally, force the direct neighbors to the coarser levels. This technique nearly remedies the spreading of
large Péclet numbers on coarse levels because it slows down the development
of newly arising large Péclet numbers.

In the case of the migration-diffusion system, we have, additionally, investigated the techniques AS and SMO, see Section 6.4 for further details.
We have observed that these techniques do not improve the convergence
of the approach since their convergence factor is bounded by the slow convergence of the smoother for the critical points in case of SMO, and by
the outer convergence in the case of AS. The same can be observed for the
MITReM. Hence, we do not display any results of these methods, here.

Summarizing, PAMG(NO_3^-,max,NO_3^-) provides better results than P-
AMG(max) for systems from early Newton steps, where the current approximation is far away from the solution. However, with better accuracy of the

approximation convergence problems arise for both approaches caused by points violating the Péclet condition. This can be remedied by using forced C-points.

When using forced C-points, the convergence of the methods PAMG(NO_3^-,max,NO_3^-) and PAMG(max) is nearly the same because the critical regions (especially the points near the electrodes) are forced to exist on all levels. Hence, the coarsening of the two methods in this part of the domain is nearly the same.

Ion-System 2

In order to investigate the effects of a physics-aware coarsening for ion-system 2, we use the 2nd (I2-M2) and the 10th (I2-M10) linear system from the Newton method. For ion-system 2, the Newton method takes 19 iterations to converge. Hence, the 2nd linear system is far away from the final system, whereas the 10th system is near the final system. This allows us to analyze the effect of a physics-aware primary matrix for linear systems from early and late Newton steps.

Ion-system 2, alike as ion-system 1, includes no homogeneous reactions. Hence, the overall behavior of ion-system 2 is similar to ion-system 1. For ion-system 2, we observe that the migration at the electrodes is underrepresented using a maximum-norm based primary matrix for linear systems from early Newton steps, too. We demonstrate that the physics-based coarsening helps to improve the convergence, especially, for system I2-M2.

approach			levels					
			2	3	4	5	6	7
PAMG(max)			0.15	0.18	0.33	0.51	0.54	div.
PAMG(Ag^+	max	Ag^+)	0.15	0.39	0.52	div.	0.92	0.90
PAMG(NO_3^-	max	NO_3^-)	0.15	0.17	0.33	0.63	0.95	0.98
PAMG(K^+	max	K^+)	0.15	0.17	0.33	div.	0.55	0.54

Table 7.16: Convergence factors for system I2-M2.

Table 7.16 shows the convergence results for the PAMG approaches considered. The divergence of the 5-level approaches PAMG(K^+,max,K^+) and PAMG(Ag^+,max,Ag^+) is caused by the direct solver on the coarsest level. In these cases, the reduction of the residual for the coarsest level system is 5 magnitudes lower than in the other cases. The coarsest level solver is investigated in more detail in Section 7.3.1. Note, that we do not use the

automatic switching to a different coarsest level solver for the tests in this
section.

We observe that the PAMG(max) method diverges when using 7 levels for system I2-M2. The PAMG approaches using an ion-concentration
primary unknown at the electrodes do not show divergence for the 7-level
approach. Further investigations show that the divergence in the case of
PAMG(max) is caused by an underrepresented migration at the electrodes
similar to the case of ion-system 1.

approach			levels					
			2	3	4	5	6	7
PAMG(max)			0.15	0.18	0.35	0.49	0.60	0.72
PAMG(Ag^+	max	Ag^+)	0.15	0.49	0.66	0.73	div.	div.
PAMG(NO_3^-	max	NO_3^-)	0.15	0.18	0.36	0.49	div.	div.
PAMG(K^+	max	K^+)	0.15	0.18	0.36	0.47	0.70	0.70

Table 7.17: Convergence factors for system I2-M10.

With exception of the 5-level method, PAMG(K^+,max,K^+) shows a
better convergence than the other PAMG(ion,max,ion) approaches considered for system I2-M2. This is also true considering system I2-M10, shown
in Table 7.17. The reason is the larger diffusion coefficient of K^+, cf. Table 7.2. This causes a larger weight of the migration in the primary matrix,
since the mechanical mobility depends on the diffusion coefficient.

Generally speaking, the ion-concentration primary unknown near the
electrodes performs best which product of diffusion coefficient an load has
the largest absolute value. This has also been observed in the case of the
migration-diffusion system.

Similar to ion-system 1, the Péclet numbers increase the better we approach the final solution. This causes divergence of the PAMG(Ag^+,max,
Ag^+) and PAMG(NO_3^-,max,NO_3^-) approaches and a higher level-dependency of all approaches considered.

Summarizing, PAMG(K^+,max,K^+) provides the best results for systems
from early Newton steps, where the current approximation is far away from
the solution. However, we have seen that this method diverges because of an
insufficiently working coarsest level solver when using 5 levels. For system
I2-M10, its results are very similar to PAMG(max).

Similar to the case of ion-system 1, the use of forced C-points helps
to improve the convergence factors of the approaches because it restricts
the number of levels performed and prevents increasing Péclet numbers on

approach				levels					
				2	3	4	5	6	7
I2-M2									
PAMG(max)-FC(1)				0.12	0.14	0.32	-	-	-
PAMG(Ag^+	max	Ag^+)-FC(1)		0.12	0.15	0.33	-	-	-
PAMG(NO_3^-	max	NO_3^-)-FC(1)		0.12	0.14	0.29	-	-	-
PAMG(K^+	max	K^+)-FC(1)		0.12	0.14	0.28	-	-	-
I2-M10									
PAMG(max)-FC(1)				0.12	0.20	0.30	-	-	-
PAMG(Ag^+	max	Ag^+)-FC(1)		0.12	0.21	0.32	-	-	-
PAMG(NO_3^-	max	NO_3^-)-FC(1)		0.12	0.22	0.32	-	-	-
PAMG(K^+	max	K^+)-FC(1)		0.12	0.21	0.31	-	-	-

Table 7.18: Convergence factors for systems I2-M2 and I2-M10.

coarse levels. However, considering the 4-level variants only, the PAMG(*)-FC(0) methods show worse results than the original approach. Hence, they are not displayed, here.

Table 7.18 shows the convergence results for the PAMG(*)-FC(1) approaches which forces all critical points and their direct neighbors to exist on all coarse levels. PAMG(*)-FC(1) shows better results than the original approach, it restricts the number of levels to 4. The differences between the PAMG(*)-FC(1) approaches are only marginal. This is caused by the fact that the coarsening at the electrodes for the approaches is nearly the same because of the usage of forced C-points as we already pointed out for ion-system 1.

Ion-System 3

We investigate the effects of a physics-aware coarsening for ion-system 3 which features homogeneous reactions and compare the results to the case without homogeneous reactions. The 2nd, 5th, and the 10th linear system from the Newton method for the channel geometry are chosen for analysis. The Newton method takes 13 iterations to converge for this ion-system. Hence, the 2nd (I3-M2) linear system is far away from the final one, the 5th linear system (I3-M5) already shows the main character of the final system, and the 10th system (I3-M10) is nearly alike the final system.

The coupling structure of the matrix for ion-system 3 is totally different to the other systems because of the homogeneous reactions, which cause

from unknown	to unknown	share	share without hom. reac.
$c(S_2O_3^{2-})$	$c(AgS_2O_3^-)$	75%	0%
$c(AgS_2O_3^-)$	$c(AgS_2O_3^-)$	14%	0%
$c(Na^+)$	c(potential)	10%	85%
$c(S_2O_3^{2-})$	$c(S_2O_3^{2-})$	0%	5%
$c(NO_3^-)$	$c(NO_3^-)$	0%	4%
c(potential)	$c(Ag(S_2O_3)_2^{3-})$	0%	5%

Table 7.19: Share for system I3-M10.

strong couplings. The reaction of this system is

$$S_2O_3^{2-} + AgS_2O_3^- \rightleftharpoons Ag(S_2O_3)_2^{3-}. \tag{7.4}$$

The concentration-to-concentration couplings of the ions involved in the homogeneous reactions dominate the maximum-norm based primary matrix. Furthermore, the coupling-structure is nearly independent of the Newton step considered. Table 7.19 shows the numbers of the coupling strength for system I3-M10 and compares it to the same system ignoring the couplings caused by homogeneous reactions. We observe that the coupling strength between the 2 reactants is very large (75%). Ignoring the homogeneous reactions, the coupling of Na^+-concentration to the potential unknown dominates which is typical for systems without homogeneous reactions.

We have not investigated the effect of homogeneous reactions in the case of the migration-diffusion model considered in Chapter 6. However, as long as the ions do not contribute to the homogeneous reactions the results are very similar. Namely, the approaches PAMG(A,max,C) perform better than PAMG(A,B,C) with B $\neq$ max. The reason is that the diffusive character is underrepresented in Ω_2 in the case of B $\neq$ max. Note again that this is only valid if the ions considered do not contribute to the homogeneous reactions. In the other case, the convergence factors are worse because the reaction usually dominates and prevents a coarsening with respect to the migration.

PAMG(NO_3^-,max,NO_3^-)'s convergence factors are in most cases slightly better than in the case of PAMG(Na^+,max,Na^+). The convergence factors for PAMG($NaS_2O_3^-$,max,$NaS_2O_3^-$) are worse than for PAMG(Na^+,max, Na^+). This corresponds to the size of the diffusion coefficients of the ions. Namely, NO_3^- has the largest one followed by Na^+ and $NaS_2O_3^-$, cf. Table 7.2. To ensure a clear representation, the following tables only show the results for PAMG(NO_3^-,max,NO_3^-) and PAMG(max) and their variant using forced C-points (FC).

approach	levels					
	2	3	4	5	6	7
PAMG(max)	0.68	0.69	0.71	0.82	0.95	-
PAMG(NO_3^-,max,NO_3^-)	0.27	0.30	0.37	0.65	0.95	0.95
PAMG(max)-FC(0)	0.76	0.76	0.77	0.80	0.92	-
PAMG(NO_3^-,max,NO_3^-)-FC(0)	0.24	0.27	0.35	0.50	0.80	-
PAMG(max)-FC(1)	0.24	0.29	0.47	0.53	0.89	-
PAMG(NO_3^-,max,NO_3^-)-FC(1)	0.23	0.30	0.47	0.54	0.86	-

Table 7.20: Convergence factors for system I3-M2.

Table 7.20 shows the convergence results for system I3-M2. We observe the benefit of the physics-based coarsening only when performing less than 6 levels for the PAMG(*) and PAMG(*)-FC(0) methods. For the PAMG(*)-FC(1) methods, the convergence factors are very similar because of the use of the forced C-points, which results in a very similar coarsening near the electrodes.

Starting from level 6 all approaches show a very bad convergence behavior. This is caused by problems of the coarsest level solver, which we have already pointed out in Section 7.3.1. Note again that we do not use the automatic switching to a different coarsest level solver for the tests in this section.

approach	levels					
	2	3	4	5	6	7
PAMG(max)	0.28	0.30	0.36	0.62	0.93	-
PAMG(NO_3^-,max,NO_3^-)	0.27	0.30	0.37	0.56	0.91	0.91
PAMG(max)-FC(0)	0.23	0.28	0.40	0.50	0.62	-
PAMG(NO_3^-,max,NO_3^-)-FC(0)	0.23	0.28	0.40	0.50	0.59	-
PAMG(max)-FC(1)	0.19	0.28	0.44	0.60	0.63	-
PAMG(NO_3^-,max,NO_3^-)-FC(1)	0.19	0.28	0.43	0.58	0.64	-

Table 7.21: Convergence factors for system I3-M5.

Comparing the convergence of PAMG(max) between systems I3-M2 and I3-M5, the convergence factor drastically improves if considering less than 6 levels. This is expected, since the electromagnetic field at this stage is already approximated reasonably well which leads to a good representation of the migration near the electrodes also for the case of a maximum-norm based primary matrix.

For system I3-M5, the convergence factors between the 5 and the 6-level PAMG(*) approaches considered drop drastically, too, see Table 7.21. However, the reason is different to system I3-M2. To be more specific, the smoother shows an oscillating behavior similar to the case of system I1-M30, see Figure 7.12. Using PAMG(*)-FC(0) this drop is reduced because the Péclet numbers on coarse levels are lower as in the case of the PAMG(*) methods. However, additionally forcing the neighbors to the coarser levels does not improve the convergence.

approach	levels					
	2	3	4	5	6	7
PAMG(max)	0.28	0.30	0.38	0.43	0.67	-
PAMG(NO_3^-,max,NO_3^-)	0.28	0.30	0.38	0.45	0.60	0.74
PAMG(max)-FC(0)	0.22	0.28	0.41	-	-	-
PAMG(NO_3^-,max,NO_3^-)-FC(0)	0.22	0.28	0.43	-	-	-
PAMG(max)-FC(1)	0.18	0.27	0.43	-	-	-
PAMG(NO_3^-,max,NO_3^-)-FC(1)	0.18	0.27	0.44	-	-	-

Table 7.22: Convergence factors for system I3-M10.

For the system I3-M10, the Péclet numbers are significantly larger than in the case of systems I3-M2 and I3-M5. This is in line with the observations for the other ion systems. Table 7.22 shows the convergence factors for this system. We see that using FC-points the convergence factor of the 4-level approach does not improve. The only benefit of the FC methods is that it restricts the number of levels. The reason for the rather poor performance of the FC-methods are the homogeneous reactions which lower the negative effect of a violated Péclet condition.

Summarizing, the convergence behavior of the approaches considered for the case of systems including homogeneous reactions is different to systems without homogeneous reactions. We do not see a benefit of the FC methods, besides the restriction of the number of levels. In some cases also FC(0) is better than FC(1). We conclude, that the problems caused by large Péclet numbers occur at a later level because of the dominance of the homogeneous reactions. Hence, the plain level-restricted PAMG(*) approaches show better results as the ones using forced C-points.

7.4   PAMG for the MITReM

We apply the physics-aware PAMG approaches developed for full simulation runs of the steady-state MITReM simulator and compare the performance of the approaches developed to the state-of-the-art PAMG approaches.

Section 7.4.1 summarizes the important aspects when defining a PAMG strategy for the MITReM and introduces the PAMG approaches considered. The approaches are compared on a range of different geometries and ion-systems described in Section 7.1. We start with the description of the results for the channel geometry in Section 7.4.2. In Section 7.4.3, we present the results of the slightly more complicated backward facing step geometry. Then, we show results for the numerically hardest and industrially most relevant crevice geometry in Section 7.4.4. Finally, we summarize all results in Section 7.4.5.

7.4.1   The PAMG Approaches Considered

We apply six different PAMG approaches to each geometry and ion-system configuration. The PAMG approaches are

- PAMGn,

- PAMGp,

- PAMGa,

- PAMGn-FC(1), and

- PAMGa-FC(1).

- PAMGn-5,

The BGS smoother does not work for the problems considered, because of the locally violated Péclet condition, cf. Section 7.3.2. Hence, all PAMG approaches make use of Block-ILU (BILU) smoothing. In order to create a suited permutation for the BILU smoother, we use the reordering framework. The RCM sorting algorithm is used, with a starting point which is chosen with the help of weights. In this way, the first point of the permuted system is always at the inflow.

Furthermore, the approaches use BICGSTAB acceleration and a direct solver as their coarsest level solver. The standard coarsest level solver employed is PARDISO. However, in the case where PARDISO does not solve the coarsest level system sufficiently well, MUMPS is used, see Section 7.3.1 for further details. The coarsening and interpolation employed in the PAMG methods is based on primary matrices. Single-unknown interpolation is used based on the same primary matrix as it is used for the coarsening. The methods differ in their choice of their coarse-grid correction. Some of the methods use a physics-aware coarsening. The coarse-grid correction and their properties are described below.

PAMGn and PAMGp

The approach PAMGn uses a straightforward primary matrix based on the maximum-norm. PAMGp uses an ion-concentration unknown based primary matrix. We use the concentration coupling of Ag^+ for systems 1 and 2 and the concentration coupling of $NaS_2O_3^-$ for system 3. The approaches have already been introduced in Section 7.2.1.

PAMGa

The PAMGa approach uses a physics-aware primary matrix. To be more specific, during the computation of the primary matrix it takes the position of the electrodes into account. We have demonstrated that the migration direction is insufficiently represented in a maximum-norm based primary matrix near the electrodes. This is caused by the nonlinearity of the system and the initial guess of a zero potential gradient. Hence, the convective character of the migration term is not present at all in the first linear system, and increases starting with the second linear system. Consequently, it is not represented well for systems from early Newton steps.

The physics-aware primary matrix stresses the migration in the primary matrix because it ignores the reactive and diffusive terms which are not present in the diagonal block (when considering an unknown-based order of the matrix). We have shown that this strategy leads to better convergence factors than the state-of-the-art choices of the primary matrix for the migration-diffusion system as well as for selected matrices from the MITReM, cf. Chapter 6 and Section 7.3.3, respectively.

We partition the domain into several parts. The parts are either around an electrode or without an electrode. If a part of the domain includes no electrode, the contained points use the maximum-norm based primary matrix. If a partition contains an electrode, an ion-concentration will be used as primary unknown. The choice of an optimal partitioning depends on the geometry, positions of the electrodes, grid resolution, and ion-system as well as the potential difference between the electrodes, and the velocity field. Hence, the descriptions of the partitioning of the domain can be found in the respective subsections for the channel, crevice and BFS geometry.

The choice of the ion-concentration unknown for the parts of the domain near the electrodes differs for each ion-system and is shown in Table 7.23. We have demonstrated that the displayed configurations show the best convergence behavior in Section 7.3.3. The reason is that the migration is more dominant if a ion has a large diffusion coefficient, which results also in a very large mechanical mobility. Furthermore, the respective ions shall not take part in the homogeneous reactions. The reason being that the reactive terms are usually dominating, so that the migration would be underrepresented.

ion-system	ion
1	$c(NO_3^-)$
2	$c(K^+)$
3	$c(NO_3^-)$

Table 7.23: Choice of ion-concentration unknown in primary matrix for the ion-systems.

PAMGn-FC(1) and PAMGa-FC(1)

In Chapter 5, we have developed the heuristic Péclet number. It can be used to locate numerical difficulties on every level. If the Péclet number at a point is larger than 2, i.e. violates the Péclet condition, we call this point critical. The PAMG approaches using forced C-points, PAMGn-FC(1) and PAMGa-FC(1), force all critical points and, additionally, all direct neighbors of critical points to exist on all coarser levels. We have demonstrated that this avoids drastic increasing Péclet numbers, and a spreading of the area with critical points on coarser levels.

Note that the computation of the Péclet numbers leads to higher setup costs of the methods. The Galerkin product has to be computed four times instead of once for each level. Since this is the computationally most intensive part of the setup, the setup cost will increase by a factor of 2 to 3. However, the Péclet numbers do not have to be computed for every Newton step, especially, for late Newton steps the Péclet numbers do not change drastically. Note that the Péclet numbers do not have to be computed for the first ion-system, because the convective part of the migration is zero in this case.

PAMGn-5

The approach PAMGn-5 uses a primary matrix based on the maximum-norm. In contrast to PAMGn its number of levels to be performed is restricted. The restriction of the number of levels is done because we observe that the convergence rate is level-dependent. The level-dependency is caused by the local violation of the Péclet condition and leads to a diverging PAMGn approach for complicated geometries. We restrict the number of levels to 5 because this is the level where the PAMG-FC methods usually stop their coarsening.

7.4.2   Channel

For all channel geometries considered, we use the ion-concentration primary unknown in a rectangle around the electrode boundaries considering PAMGa(-FC(1)). The sub-domains around the electrodes have a height of $2r$ and a length of $l_e + 2r$, where l_e is the electrode length and $r = 0.00002$.

approach	ion-system 1	2	3
PAMGn-FC(1)	**106**	**101**	85
PAMGa-FC(1)	111	104	83
PAMGn	227	158	115
PAMGa	212	183	119
PAMGp	495	509	381
PAMGn-5	207	134	**79**

Table 7.24: Linear iteration counts for CHANNEL-SMALL.

Table 7.24 shows the iteration counts for solving the MITReM for the CHANNEL-SMALL geometry using the ion-systems considered. We observe that the approaches employing forced C-points perform the lowest number of iterations with exception of ion-system 3. This is what we expect when considering the convergence results for single matrices. The reason for relatively low iteration counts and, especially, the good performance of PAMGn-5 for ion-system 3 are the homogeneous reactions which lower the problems caused by a locally violated Péclet condition, cf. Section 7.3.3.

Overall, the approaches PAMGn-FC(1) and PAMGa-FC(1) are working well. They differ only slightly in their iteration counts because the coarsening near the electrodes is nearly the same. The reason being, the use of forced C-Points.

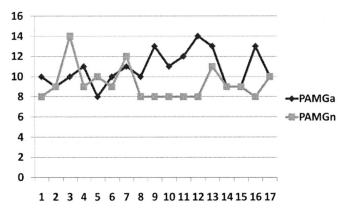

Figure 7.15: Linear iterations per system for CHANNEL-SMALL using ion-system 2.

From the experiments for the single matrices, we expect that PAMGn works better than PAMGa for linear systems from late Newton steps. In contrast, PAMGa is expected to work better than PAMGn for the linear systems from early Newton steps. Figure 7.15 shows the number of iterations needed to solve each linear system from the Newton method for ion-system 2. Indeed, for early systems, PAMGa is overall slightly better and for late systems PAMGn turns out to be better. This is also seen for ion-systems 1 and 3. However, in the latter case, it is not that significant because of the homogeneous reactions.

PAMGp shows the largest iteration counts of all methods. This is caused by the insufficient coarse-grid correction of points in the interior of the domain, as we have observed in the single matrix test for the migration-diffusion system, cf. Section 6.2.

	ion-system		
approach	1	2	3
PAMGn-FC(1)	151	**124**	116
PAMGa-FC(1)	**147**	130	112
PAMGn	293	228	165
PAMGa	290	267	166
PAMGp	628	646	315
PAMGn-5	271	190	**107**

Table 7.25: Linear iteration counts for CHANNEL-BIG.

Table 7.25 shows the iteration counts for CHANNEL-BIG. We observe

that the quality of the results is very similar to CHANNEL-SMALL. However, the iteration counts are larger. This points to problems with the
scalability of the methods.

The reason for the poor scalability is that the number of points where the
Péclet condition is violated is larger for CHANNEL-BIG than for CHAN-
NEL-SMALL. This is caused by the refinement of the grid which is not
fine enough to prevent a violation of the Péclet condition. Hence, the convergence behavior of the methods worsens. As a consequence, a drastic
increase of the iteration counts can be observed for PAMGn-5 method for
ion-systems 1 and 2 because the level-dependency of the convergence factor
increases. For ion-system 3, this effect is not that significant because of the
damping effect of the homogeneous reactions.

7.4.3   Backward Facing-Step

The backward facing step geometry has areas of recirculating flow. Hence,
the convection has a different character than in the channel geometries.
Besides of this, the geometries are very similar.

For PAMGa and PAMGa-FC(1) we use a different partitioning of the
domain than in the case of the channel. To be more specific, the size
of the sub-domains near the electrodes than in the case of the channel
geometry because of the different velocity fields. The value of r is changed
to $r = 0.00005$.

approach	ion-system		
	1	2	3
PAMGn-FC(1)	112	109	77
PAMGa-FC(1)	**108**	**106**	**68**
PAMGn	200	364	191
PAMGa	(9)	316	341
PAMGp	489	557	355
PAMGn-5	171	176	73

Table 7.26: Linear iteration counts for BFS-SMALL. Numbers in brackets
denote the iteration for which the PAMG approach does not solve accurate
enough/diverges.

Table 7.26 shows the iteration counts for the BFS-SMALL geometry.
The character of the results is very similar to the one of the channel geometry. Namely, we observe a very good performance of the methods using
forced C-points for ion-systems 1 and 2. Here, the problems caused by

the local violation of the Péclet condition lead to relatively large iteration counts for the other methods. PAMGa even diverges for the 9th linear system of ion-system 1. The divergence is caused by the large Péclet numbers at this stage which spread very fast during the coarsening so that they are not captured well enough by this approach. We have observed a similar behavior in the case of the migration-diffusion equation in Section 6.3.

The large Péclet numbers are also the reason for the relative bad performance of the PAMGn-5 method compared to the channel geometry for ion-system 2. In the case of ion-system 3, the problems caused by a locally violated Péclet condition occur at a later stage in the coarsening process because of the homogeneous reactions. Hence, the PAMGn-5 method performs similar to the methods using forced C-points. This has also been observed in the case of the channel geometry.

		ion-system	
approach	1	2	3
PAMGn-FC(1)	**271**	286	1042
PAMGa-FC(1)	273	**221**	911
PAMGn	(8)	574	(3)
PAMGa	344	361	(4)
PAMGp	609	652	578
PAMGn-5	323	220	230

Table 7.27: Linear iteration counts for BFS-BIG.

Table 7.27 shows the iteration counts for the BFS-BIG geometry. Similar to the case of the channel geometry, the higher refinement leads to more points which violate the Péclet condition. This causes larger iteration counts for all methods.

We observe that PAMGa diverges for ion-systems 3. This is caused by an insufficient solution of the coarsest level solver. Also the performance for the other methods using forced C-points for ion-system 3 is very bad. We have already seen that the usage of forced C-points for the other geometries, and also in the case of the single matrix tests does not give any benefit besides the restriction of the number of levels. For the BFS-BIG geometry the usage of forced C-points effects the convergence negatively. This is caused by the coarsest level solvers which do not solve the critical points on the coarsest level accurate enough, cf. discussion of the coarsest level solver in Section 7.3.1.

7.4.4   Crevice

We already pointed out that the first guess is very far away from the solution for ion-system 1 which causes a very large number of Newton steps. In the case of the crevice geometry, this seems to be even more drastic, many of the methods considered do not give sufficiently accurate results in the course of the Newton method. Hence, they do end up with a system which they cannot solve.

In the case of the crevice geometries, PAMGa and PAMGa-FC(1) use the ion-concentration primary unknown in a rectangle around the electrode boundaries with a height of $2r$ and a length of $l_e + 2r$ for horizontal electrodes. For vertical electrodes we use the height $h_e + 2r$ and length $2r$, where h_e is the electrode height, l_e is the electrode length and $r = 0.00002$.

approach	ion-system 1	2	3
PAMGn-FC(1)	**155**	188	**93**
PAMGa-FC(1)	(40)	**182**	102
PAMGn	(6)	261	677
PAMGa	(36)	274	149
PAMGp	439	531	128
PAMGn-5	(26)	187	99

Table 7.28: Linear iteration counts for CREVICE-SMALL.

Table 7.28 shows the iteration counts of the approaches for the CREVICE-SMALL geometry. The results are similar to the other SMALL geometries. Namely, the methods using forced C-points have a significantly lower iteration count than the other approaches with exception of the PAMGn-5 method. For system 2, the PAMGn-5 method needs a relatively low number of iterations because of the restricted number of levels. This prevents a drastic impact of the locally violated Péclet condition on the convergence factor of PAMGn-5.

Table7.29 shows the iteration counts for the CREVICE-BIG geometry. For ion-system 1, the PAMGa-FC(1) approach is better than for the SMALL geometry. For ion-system 2 the PAMGn-5 method needs nearly the same number of iterations as in the case of the SMALL one. The reason for this is the crack. In the crack the grid-refinement, in contrast to the other geometries, is fine enough so that the number of large Péclet numbers in this part of the domain stays nearly constant. Nonetheless, more large Péclet

| | ion-system | | |
approach	1	2	3
PAMGn-FC(1)	**144**	216	(1)
PAMGa-FC(1)	(35)	237	128
PAMGn	(4)	532	(1)
PAMGa	(35)	299	134
PAMGp	547	687	171
PAMGn-5	(37)	**189**	**101**

Table 7.29: Linear iteration counts for CREVICE-BIG.

numbers occur at the upper electrode, hence, the FC methods are a bit worse for ion-system 2.

Considering ion-system 3, we observe that PAMGn(-FC(1)) diverges for the first linear system. This is caused by a bad coarse-grid correction. The reason being a bad representation of the migration in the coarsening process.

7.4.5 Summary

The full runs confirm the findings from the single matrix tests. Namely, for ion-systems without homogeneous reactions it is beneficial to make use of forced C-points in order to prevent a strong level-dependency of the convergence. For the ion-system which makes use of homogeneous reactions, the level-dependency caused by a locally violated Péclet condition is damped. Hence, a restriction of the number of levels to be performed in dependence of the Péclet numbers on coarse levels is sufficient to achieve good results. Furthermore, using forced C-points, in this case, does not pay-off since the coarsest level solver is not able to solve critical points sufficiently well.

Overall, PAMGn-FC(1) shows the best results for ion-systems 1 and 2. The reason being that it uses nearly the same coarsening strategy as PAMGa at the electrodes for linear systems from early Newton steps because of the use of forced C-points. Furthermore, the forced C-points remedy the arising problems caused by a violated Péclet condition.

PAMGn-5, which uses a restriction of the number of levels, shows the best results for ion-system 3. The approach restricts the number of levels to 5. The number of levels to be performed can be further optimized by using the heuristic Péclet number. The approach stops the coarsening if the number of critical points exceeds a certain number of the overall points. Concluding, for productive runs, we suggest the use of PAMGn-FC(1) for ion-systems 1 and 2 and PAMGn-5 for ion-system 3.

We have also observed that the choice of a physics-aware primary matrix leads to better performance than the state-of-the-art approach for linear systems from early Newton steps. However, for late Newton steps this strategy is usually counter productive because the migration term is, then, given too much influence in the primary matrix. This can be circumvented if analyzing the local violations of the Péclet condition. We suggest that the physics-aware PAMG approach switches to a purely maximum-norm based primary matrix if the Péclet numbers near the electrodes exceed a certain barrier.

Note that the run time per cycle of the physics-aware approaches is usually longer than for the straight forward approaches. The reason being that the Péclet numbers on all levels have to be computed. This leads to a 2-3 times higher setup cost, however, this is overcompensated by far by the drastically lower iteration counts compared to PAMGn and PAMGp.

7.5 Conclusion

We have demonstrated that the PAMG technology allows the creation of linear solvers which outperform the standard ILU-GMRES one-level solver of the MITReM simulator by a factor of 2 to 3 considering whole simulation runs.

Furthermore, we have identified problems occurring when applying state-of-the-art PAMG to the MITReM. The problems result in bad convergence factors, or even divergence for the straightforward PAMG approach (PAMGn).

State-of-the-art PAMG approaches do only work with the linear system considered and usually do not use further information. We have developed several techniques to incorporate physical properties of the underlying PDE system into PAMG approaches. The physic-aware PAMG framework newly developed has been compared to state-of-the-art approaches for full simulation runs. The results confirm the benefits of the new approach. We briefly describe the benefits of the physics-aware PAMG approach for the MITReM.

In the case of the MITReM, difficulties arise because of the nonlinearity of the system, the convection dominance in some part of the domain, and the migration dominance in other parts.

We have demonstrated that the physics-based reduction techniques developed and incorporated in the reordering framework introduced, enable a much better representation of the dominant flow in the ordering of the variables of the linear system. We have shown that these flow-based orderings lead to much better smoothing results than state-of-the-art reordering techniques for block Gauss-Seidel smoothing.

Furthermore, we have demonstrated the benefits of a physics-based primary matrix. The physics-based primary matrix allows us to strengthen the migrative character at the electrodes, which is underrepresented in a maximum-norm based primary matrix due to the zero initial guess of the potential gradient.

Finally, we have demonstrated that the choice of forced C-points based on the information of the heuristic Péclet number developed nearly remedies the level-dependency of the PAMG approaches which is caused by a locally violated Péclet condition.

Chapter 8

Conclusions and Future Work

The multi-ion transport and reaction model (MITReM) is a non-linear convection-dominated system of PDEs, which describes convective, diffusive, reactive and migrative effects. In order to solve the system, it is discretized and linearized resulting in a series of large sparse matrices. The standard solver used in the MITReM simulation software considered is an ILU preconditioned GMRES procedure. Such type of solvers show a super-linearly increasing run time with increasing problem size. We have demonstrated that the run time of a MITReM simulation can drastically be reduced if AMG technology is used as a preconditioner for Krylov methods employed for the solution of linear systems.

The complexity of the system makes it impossible to robustly *and* efficiently solve the arising linear systems with state-of-the-art multigrid strategies. The main reason for this is that physical properties of the PDE system are not represented well in the standard multigrid framework.

In this work, we have developed a physics-aware multigrid approach for both scalar and systems of PDEs with special focus on the steady-state MITReM. Physics-aware means that it incorporates explicitly given physical information of the underlying PDE (system) in its **smoothing** and **coarse-grid correction** strategy. Hence, it is able to solve the arising linear systems in a robust *and* efficient way.

It is well-known that the ordering of variables within the smoothing procedure for a convection-dominated problem like the MITReM can drastically influence the performance of classical AMG. For example, a Gauss-Seidel-type smoothing performs best using a flow-wise ordering and ILU-type smoothing performs best using a reverse Cuthill McKee ordering. That is, the best ordering for a situation is not only dependent on the physics involved, but also on the smoother employed.

We have developed a **reordering framework** to create permutations for linear systems from scalar as well as systems of PDEs which is applicable in all simulation environments where a reordering helps to improve the efficiency of the solver employed. In particular, the framework enables the creation of orderings based on additional physical information, such as vector fields describing global direction of movement. Furthermore, we have introduced a technique to compute vector fields which represent the global flow based on the linear system and the coordinate information. These vector fields accurately represent the direction of dominant flow.

Using the reordering framework, we have demonstrated that the knowledge of the direction of flow based on vector fields (convective as well as migrative) can be used to generate orderings which represent the **flow direction** much better than using the standard reordering techniques, which do not make use of such information.

The proposed physics-aware smoothing procedure uses the reordering framework to create permutations of the system's variables which enable the improvement of the smoothing behavior. We have shown the benefit of a smoother- and **physics-adapted ordering** for the convection-diffusion equation as well as for the MITReM.

Furthermore, we have demonstrated that convergence problems arise when applying the standard coarse-grid correction strategy to linear systems from the MITReM as well as from a migration-diffusion system, which can be used to simulate ionic transport processes. These problems are caused by both the **nonlinearity** and the choice of the **discretization**, i.e., local violation of the Péclet condition. Hence, besides the physics-aware smoothing, we have created a physics-aware coarse-grid correction. The physics-aware coarse-grid correction allows us to take physical properties of the PDE system into account which are not present or not represented well in the linearized systems itself.

We have demonstrated that the initial guess of the Newton linearization employed leads to an insufficient representation of the potential gradient for early Newton steps. This causes state-of-the-art PAMG approaches to show a bad convergence behavior or even divergence for linear systems from early Newton steps. The **physics-aware coarse-grid correction** developed shows much better results because it is able to stress the relevance of the migration in the coarse-grid correction process by using external information like the position of the electrodes.

In conjunction with that coarse-grid correction approach, we have developed a **heuristic Péclet number** which is **purely based on algebraic criteria**, and does not explicitly use grid information. Hence, it can also be computed on coarse levels where no grid is available. We have shown that the heuristic Péclet number is equivalent to the mesh Péclet number for the convection-diffusion equation on a regular grid. Additionally, we have generalized the heuristic Péclet number to the case of systems of PDEs in order to apply it in the case of the MITReM. The heuristic Péclet number takes the convection, migration and diffusion of the MITReM system into account.

We have investigated the effects of a locally violated Péclet condition on the hierarchy of classical AMG. The Péclet numbers increase and the area of violated Péclet condition spreads out on coarse levels when using standard AMG components.

We have introduced several physics-aware coarse-grid correction strategies in the context of **scalar PDEs as well as systems of PDEs**. The proposed strategies use the heuristic Péclet number to localize numerical difficulties in each point on each level of the AMG method. In the case of strongly changing convection/migration coefficients, we have shown that a robust coarse-grid correction can be achieved by forcing variables/points which violate the Péclet condition to the coarsest level where they can be solved by a direct solver.

The approach proposed to solve the MITReM makes use of the physics-aware AMG approach newly developed. The PAMG method uses the reordering framework to improve the smoothing and the heuristic Péclet number to improve the coarse-grid correction. We have motivated the use of the physics-based PAMG components and applied the approach for a range of geometries and ion-systems with high importance to industry. Our numerical experiments recommend, in particular, the explicit consideration of physical properties of the underlying problem for constructing AMG-based preconditioners to ensure a robust and efficient PAMG approach.

Bringing all of the individual aspects together provides an efficient and robust new variant of the AMG approach for applications for which standard variants are no longer effective. The effectiveness of this new approach has been demonstrated for linear systems from the MITReM, convection-diffusion equations, and migration-diffusion systems. The new components have been compared to state-of-the-art classical AMG on a range of model problems to show their robustness and efficiency.

Future Work

We have shown that the physics-aware AMG approach newly developed can be used to solve linear systems from the MITReM, convection-diffusion equations and migration-diffusion systems, and show more robustness than standard AMG components. Future work includes the application of this approach to other industrially relevant applications such as semiconductor process and device simulation, oil-reservoir, and reactor simulation.

Furthermore, there are several possibilities to improve the MITReM simulation software employed. The Péclet condition is locally violated which is caused by the discretization-type of the migration term. Hence, it is worth analyzing if the finite element discretization of the migration term can be replaced by a residual-distribution discretization using the N-Scheme as it is already used for the convective term. Especially, it has to be verified that this change in the discretization does not have a negative influence on the results near the electrode boundaries.

In order to speed up the Newton method and enhance its robustness, two aspects have to be improved. First of all, the first guess of the Newton method has to be improved. This is especially true for special model-setups where only two ions are involved or where the limiting current is approached. This can be achieved either numerically by using bisection, or by determining a better suited first guess. Secondly, the static-damping strategy should be replaced by an adaptive one to enhance the robustness of the Newton method when approaching the limiting current.

Bibliography

[1] P. R. Amestoy, I. S. Duff, and J.-Y. L'Excellent. A fully asynchronous multifrontal solver using distributed dynamic scheduling. *SIAM Journal on Matrix Analysis and Applications*, 23(1):15–41, 2001.

[2] P.W. Atkins and J. de Paula. *Physical Chemistry*. Oxford University Press, 9th edition, 2010.

[3] E. Bank, J. W. Wan, and Z. Qu. Kernel preserving multigrid methods for convection-diffusion equations. *SIMAX*, 27(4):1150–1171, 2006.

[4] T. Bányai, D. V. Abeele, and H. Deconinck. A fast fully-coupled solution algorithm for the unsteady incompressible Navier-Stokes equations. Conference on Modelling Fluid Flow (CMFF), The 13th International Conference on Fluid Flow Technologies. Hungary, 2006.

[5] J. Bey and G. Wittum. Downwind numbering: robust multigrid for convection-diffusion problems. *Appl. Numer. Math.*, 23(1):177–192, 1997.

[6] R. Blaheta. A multilevel method with overcorrection by aggregation for solving discrete elliptic problems. *J. Comp. Appl. Math.*, 24:227–238, 1988.

[7] J.O.M. Bockris and A.K.N. Reddy. *Modern electrochemistry*. Plenum, New York, 1977.

[8] L. Bortels. *The multi-dimensional upwinding method as a simulation tool for the analysis of multi-ion electrolytes controlled by diffusion, convection and migration*. PhD thesis, Vrije Universiteit Brussel, Belgium, 1996.

[9] L. Bortels, J. Deconinck, and B. van den Bossche. The multi-dimensional upwinding method (MDUM) as a new simulation tool for the analysis of multi-ion electrolytes controlled by diffusion, convection and migration. *J. Electroanal. Chem.*, 404:15–26, 1996.

[10] L. Bortels, B. van den Bossche, J. Deconinck, S. Vandeputte, and A. Hubin. Analytical solution for the steady-state diffusion and migration involving multiple reaction ions. Application to the identification of Butler-Volmer kinetic parameters for the ferri-/ferrocyanide redox couple. *Journal of Electroanalytical Chemistry*, 429(1-2):139–155, 1997.

[11] D. Braess. Towards algebraic multigrid for elliptic problems of second order. *Computing*, 55:379–393, 1995.

[12] D. Braess. *Finite Elements*. Cambridge University Press, Cambridge, 2nd edition, 2001.

[13] A. Brandt. Algebraic multigrid theory: The symmetric case. *Appl. Math. Comp.*, 19:23–56, 1986.

[14] A. Brandt. General highly accurate algebraic coarsening. *ETNA*, 10:1–20, 2000. Special issue on the Ninth Copper Mountain Conference on Multilevel Methods.

[15] A. Brandt. Multiscale and multiresolution methods: Theory and application. In T.J. Barth, T.F. Chan, and R. Haimes, editors, *Multiscale scientific computation: review*, pages 1–96. Springer Verlag, Heidelberg, 2001.

[16] A. Brandt, S. McCormick, and J. Ruge. Algebraic multigrid (AMG) for automatic multigrid solution with application to geodetic computations. Technical Report POB 1852, Institute for Computational Studies, Fort Collins (CO), USA, 1982.

[17] A. Brandt, S. McCormick, and J. Ruge. Algebraic multigrid (AMG) for sparse matrix equations. *Sparsity and its Applications*, pages 257–284, 1984. D. Evans, Ed. Cambridge University Press.

[18] A. Brandt, J. Brannick, K. Kahl, and I. Livshits. Bootstrap AMG. *SIAM J. on Scientific Computing*, 33(2):612–632, 2011. doi: 10.1137/090752973.

[19] J. Brannick and L. Zikatanov. Algebraic multigrid methods based on compatible relaxation and energy minimization. In O. B. Widlund and D. E. Keyes, editors, *Domain Decomposition Methods in Science and Engineering XVI*, volume 55 of *Lecture Notes in Computational Science and Engineering*, pages 15–26. Springer Berlin Heidelberg, 2007.

[20] J. Brannick, M. Brezina, D. Keyes, O. Livne, I. Livshits, S. MacLachlan, T. Manteuffel, S. McCormick, J. Ruge, and L. Zikatanov. Adaptive smoothed aggregation in lattice QCD. In O. B. Widlund and D. E. Keyes, editors, *Domain Decomposition Methods in Science and Engineering XVI*,

volume 55 of *Lecture Notes in Computational Science and Engineering*, pages 505–512. Springer Berlin Heidelberg, 2007.

[21] M. Brezina, A. Clearly, R. Falgout, V. Henson, J. Jones, T. Manteuffel, S. McCormick, and J. Ruge. Algebraic multigrid based on element interpolation (AMGe). *SIAM J. Sci. Comp.*, 22:1570–1592, 2000.

[22] M. Brezina, R. Falgout, S. MacLachlan, T. Manteuffel, S. McCormick, and J. Ruge. Adaptive smoothed aggregation (αSA). *SIAM Rev.*, 47:317–346, 2005.

[23] F. Brezzi and M. Fortin. Mixed and hybrid finite element methods. *Comp. Math.*, 15, 1991.

[24] G. F. Carey and A. Pardhanani. Multigrid solution and grid redistribution for convection-diffusion. *Internat. J. Numer. Methods Engrg.*, 27:655–664, 1989.

[25] T. Chan, J. Xu, and L. Zikatanov. An agglomeration multigrid method for unstructured grids. In J. Mandel, C. Farhat, and X.-C. Cai, editors, *Contemporary Mathematics*, volume 218, pages 67–81, 1998.

[26] W. M. Chan and A. George. A linear time implementation of the reverse cuthill-mckee algorithm. *BIT*, 20(1):8–14, 1980.

[27] T. Chartier, R. Falgout, V. Henson, J. Jones, T. Manteuffel, S. McCormick, J. Ruge, and P. S. Vassilevski. Spectral amge (rho-amge). *SIAM J. Sci. Comp.*, 25-1:1–26, 2003.

[28] P.G. Ciarlet. *The finite element method for elliptic problems*. Studies in mathematics and its applications. North-Holland Pub. Co., 1978. ISBN 9780444850287.

[29] T. Clees. *AMG Strategies for PDE Systems with Applications in Industrial Semiconductor Simulation*. PhD thesis, Universität zu Köln, 2004.

[30] B. Cockburn, G.E. Karniadakis, and C.-W. Shu. Discontinuous galerkin finite element methods. *Lecture Notes in Computational Science & Engineering, Springer-Verlag*, 11, 2000.

[31] E. Cuthill and J. McKee. Reducing the bandwidth of sparse symmetric matrices. In *Proceedings of the 1969 24th national conference*, pages 157–172, New York, NY, USA, 1969. ACM. doi: http://doi.acm.org/10.1145/800195.805928.

202

[32] C. Dan. *Contribution to the modeling of electrochemical systems: Multi-ion model in presence of gas evolution*. PhD thesis, Vrije Universiteit Brussel, Belgium, 2006.

[33] P. M. de Zeeuw and E. J. van Asselt. The convergence rate of multi-level algorithms applied to the convection-diffusion equation. *SIAM J. Sci. Stat. Comput*, 6:492–503, 1985.

[34] I. S. Duff, A. M. Erisman, and J. K. Reid. *Direct Methods for Sparse Matrices*. Oxford University Press, 1989.

[35] M. Feistauer, J. Felcman, and M. Lukáčová-Medviďová. On the convergence of a combined finite volume-finite element method for nonlinear convection-diffusion problems. *Numerical Methods for Partial Differential Equations*, 13(2):163–190, 1997.

[36] L. P. Franca and A. Russo. Deriving upwinding, mass lumping and selective reduced integration by residual-free bubbles. *Appl. Math. Lett.*, 9, 5(88): 83, 1996.

[37] *SAMG   User's   Manual.*   Fraunhofer   Institute   SCAI, www.scai.fraunhofer.de/samg.html. Version 25a1.

[38] H. Gajewski and K. Gärtner. On the discretization of van Roosbroeck's equations with magnetic field. *Z. angew. Math. Mech. (ZAMM)*, 76:247–264, 1996.

[39] A. George and J. Liu. The evolution of the minimum degree ordering algorithm. *SIAM Review*, 31(1):1–19, 1989.

[40] M. Griebel, T. Dornseifer, and T. Neunhoeffer. *Numerische Simulation in der Strömungsmechanik – eine praxisorientierte Einführung*. Vieweg, Wiesbaden, 1995.

[41] M. Griebel, T. Neunhoeffer, and H. Regler. Algebraic multigrid methods for the solution of the Navier-Stokes equations in complicated geometries. *Int. J. Num. Meth. Fluids*, 26:281–301, 1998.

[42] H. Guillard and P. Vaněk. An aggregation multigrid solver for convection-diffusion problems on unstructured meshes. In *Report 130*. University of Denver, 1998.

[43] S. Gutsch and T. Probst. Cyclic and feedback vertex set ordering for the 2D convection-diffusion equation. Technical report, Universität Kiel, 1997.

[44] G. Haase, U. Langer, S. Reitzinger, and J. Schöberl. Algebraic multigrid methods based on element preconditioning. *Int, J. Comp. Math.*, 78:575–598, 2001.

[45] W. Hackbusch and T. Probst. Downwind Gauss-Seidel smoothing for convection dominated problems. *Numerical Linear Algebra with Applications*, 4(2):85–102, 1997.

[46] W. Hackbusch, S. Gutsch, J. F. Maitre, and F. Musy. The appropriate numbering for the multigrid solution of convection dominated problems, 1998.

[47] C.A. Hall, J.C. Cavendish, and W.H. Frey. The dual variable method for solving fluid flow difference equations on Delaunay triangulations. *Computers and Fluids*, 20(2):145–162, 1991.

[48] P.W. Hemker. *Mixed defect correction iteration for the accurate solution of the convection diffusion equation*, pages 485–501. Springer, 1982.

[49] A. Henson and P. S. Vassilevski. Element-free AMGe: General algorithms for computing interpolation weights. *SIAM J. SCi. Comp.*, 23:629–650, 2001.

[50] T.J.R. Hughes. *The Finite Element Method: Linear Static and Dynamic Finite Element Analysis*. Prentice-Hall, Engelwoods Cliffs, 1987.

[51] T.J.R. Hughes, L.P. Franca, and M. Balestra. A new finite element formulation for computational fluid dynamics: V. circumventing the Babuška-Brezzi condition: A stable Petrov-Galerkin formulation of the Stokes problem accommodating equal-order interpolations. *Comput. Meths. Appl. Mech. Engrg.*, 59:85–99, 1986.

[52] T.J.R. Hughes, L.P. Franca, and M. Mallet. A new finite element formulation for computational fluid dynamics: VI. convergence analysis of the generalized SUPG formulation for linear time-dependent multi-dimensional advective-diffusive systems. *Comput. Meths. Appl. Mech. Engrg.*, 63:97–112, 1987.

[53] *Intel Math Kernel Library - Reference Manual*. Intel. URL http://software.intel.com/sites/products/documentation/doclib/mkl_sa/11/mklman/index.htm.

[54] K. Johannsen. *Robuste Mehrgitterverfahren für die Konvektions-Diffusions Gleichung mir wirbelbehafteter Konvektion*. PhD thesis, Ruprecht-Karls-Universität Heidelberg, 1998.

204

[55] C. Johnson and J. Saranen. Streamline diffusion methods for the incompressible Euler and Navier-Stokes equations. *Math. Comp.*, 47(175):1–18, 1986.

[56] J. Jones and P. S. Vassilevski. AMGe based on element agglomeration. *SIAM J. Sci. Comp.*, 23:109–133, 2001.

[57] H. Kim, J. Xu, and L. Zikatanov. A multigrid method based matching in graph for convection diffusion equations. *Num. Lin. Alg. and Appl.*, 10: 181–195, 2003.

[58] H. Kim, J. Xu, and L. Zikatanov. Uniformly convergent multigrid methods for convection–diffusion problems without any constraint on coarse grids. *Advances in Computational Mathematics*, 20(4):385–399, 2004.

[59] R. Krahl and E. Bänsch. Computational comparison between the Taylor-Hood and the conforming Crouzeix-Raviart element. *Proceedings Of ALGOMITRY*, pages 369–379, 2005.

[60] A. Krechel and K. Stüben. Operator dependent interpolation in algebraic multigrid. In *Lecture Notes in Computational Science and Engineering*, volume 3. In Proceedings of the Fifth European Multigrid Conference, Stuttgart, October 1-4, 1996, Springer, Berlin, 1998.

[61] G. Larson, D. Synder, D.V. Abeele, and T. Clees. Application of single-level, pointwise algebraic, and smoothed aggregation multigrid methods to direct numerical simulations of incompressible turbulent flows. *Computing and Visualization in Science*, 11(1):27–40, 2008. Accepted for publication.

[62] S. Le Borne. *Multigrid Methods for Convection-Dominated Problems*. PhD thesis, Christian-Albrechts-University of Kiel, 1999.

[63] S. Le Borne. Ordering techniques for two- and three-dimensional convection-dominated elliptic boundary value problems. *Computing*, 64: 123–155, 2000.

[64] D.K. Lilly. On the computational stability of numerical solutions of time dependent non-linear geophysical fluid dynamics problems. *Monthly Weather Review*, 93:11–25, 1965.

[65] O. E. Livne. Coarsening by compatible relaxation. *Numer. Linear Algebra Appl.*, 11:205–227, 2004. doi: 10.1002/nla.378.

[66] E.L. Lloyd, M.L. Soffa, and C.-C. Wang. On locating minimum feedback vertex sets. *J. Comput. Syst. Sci.*, 37(3):292–311, 1988. ISSN 0022-0000. doi: http://dx.doi.org/10.1016/0022-0000(88)90009-8.

[67] J. Mandel. Local approximation estimators for algebraic multigrid. *Electronic Transactions on Numerical Analysis*, 15:56–65, 2003.

[68] J. Mandel, M. Brezina, and P. Vaněk. Energy optimization of algebraic multigrid bases. *Computing*, 62:205–228, 1999.

[69] T. Manteuffel, S. McCormick, M. Park, and J. Ruge. Operator-based interpolation for bootstrap algebraic multigrid. *Numerical Linear Algebra with Applications*, 17(2-3):519–537, 2010. doi: 10.1002/nla.711.

[70] P.A. Markowich. *The Stationary Semiconductor Device Equations.* Springer, Wien, 1986.

[71] P.A. Markowich, C.A. Ringhofer, and C. Schmeiser. *Semiconductor Equations*. Springer, Wien, 1990.

[72] K. W. Morton. Numerical solution of convection-diffusion problems. *Applied Mathematics and Mathematical Computation*, 12, 1996.

[73] A. Napov and Y. Notay. Algebraic analysis of aggregation-based multigrid. *Numerical Linear Algebra with Applications*, 18(3):539–564, 2011. ISSN 1099-1506. doi: 10.1002/nla.741. URL http://dx.doi.org/10.1002/nla.741.

[74] A. Naumovich, M. Förster, and R. Dwight. Algebraic multigrid within defect correction for the linearized euler equations. *Numerical Linear Algebra with Applications*, 17(3-4):307–324, 2010.

[75] G. Nelissen. *Simulation of Multi-Ion Transport in Turbulent Flow*. PhD thesis, Fakulteit Toegepaste Weteneschappen, Vrije Universiteit Brussel, 2003.

[76] G. Nelissen, A. Van Theemsche, C. Dan, B. Van Den Bossche, and J. Deconinck. Multi-ion transport and reaction simulations in turbulent parallel plate flow. *J. Electroanal. Chem.*, 563:213–220, 2004.

[77] J. Newman and K. E. Thomas-Alyea. *Electrochemical Systems*. Wiley-Interscience, 3rd edition, 2004.

[78] R.A. Nicolaides. *The Covolume Approach to Computing Incompressible Flow*. Springer: Berlin/New York, 1993. Hussaini, M.Y., Kumar, A. and Salas, M.D. (eds.).

[79] F. Obermüller. Effizienzsteigerung bei AMG-methoden mit schwerpunkt auf der behandlung von Wells in der reservoirsimulation. Master's thesis, Universität zu Köln, 2012.

[80] T. Okusanya, D. L. Darmofal, and J. Peraire. Algebraic multigrid for sta-
      bilized finite element discretizations of the Navier-Stokes equations. *Com-*
      puter Methods in Applied Mechanics and Engineering, 193(33-35):3667 –
      3686, 2004. ISSN 0045-7825. doi: DOI:10.1016/j.cma.2004.01.025.

[81] M. A. Olshanskii and A. Reusken. Convergence analysis of a multigrid
      method for a convection-dominated model problem. *SIAM J. Numer.*
      Anal., 42:1261–1291, 2004.

[82] H. Paillère. *Multidimensional Upwind Residual Distribution Schemes for*
      the Euler and Navier-Stokes Equations on Unstructured Grids. PhD thesis,
      ULB, 1995.

[83] H. Paillère, H. Deconinck, and E. van den Weide. Upwinding residual
      distribution methods for compressible flow: An alternative to finite volume
      and finite element methods. *VKI LS*, 2, 1997.

[84] B. Pollul and A. Reusken. Numbering techniques for preconditioners in
      iterative solvers for compressible flows. *Journal for Numerical Methods in*
      Fluids, 55(3):241 – 261, 2006.

[85] M. J. Raw. Robustness of coupled algebraic multigrid for Navier - Stokes
      equations. In *34th AIAA*, 1996.

[86] H. Rentz-Reichert. A comparison of smoothers and numbering strategies
      for laminar flow around a cylinder. In *Flow Simulation with High Perfor-*
      mance Computers II, volume 52 of Notes on Numerical Fluid Mechanics,
      volume 52, pages 134–149, 1996.

[87] T. W. Roberts, D. Sidilkover, and R. C. Swanson. Textbook multigrid
      efficiency for the steady Euler equations. Technical report, NASA, 2004.

[88] H.-G. Roos, M. Stynes, and L. Tobiska. *Numerical Methods for Singularly*
      Perturbed Differential Equations. Springer-Verlag, Berlin, 1996.

[89] J. Ruge and K. Stüben. Efficient solution of finite difference and finite
      element equations by algebraic multigrid (AMG). Technical Report 98,
      GMD, 1984.

[90] J. Ruge and K. Stüben. Algebraic multigrid (AMG). *Frontiers in Applied*
      Mathematics, 3:73–130, 1987.

[91] Y. Saad. *Iterative Methods for Sparse Linear Systems*. SIAM, 2nd edition,
      2003. URL http://www-users.cs.umn.edu/ssaad/books.html.

[92]   O. Schenk. *Scalable Parallel Sparse LU Factorization Methods on Shared Memory Multiprocessors.* PhD thesis, ETH Zurich, 2000.

[93]   M. Schwartz. *Principles of Electrodynamics.* Dover Pubn Inc., 1987.

[94]   S. Selberherr. *Analysis and Simulation of Semiconductor Devices.* Springer, Wien, 1984.

[95]   A.V. Sokirko and F.H. Bark. Diffusion-migration transport in a system with Butler-Volmer kinetics, an exact solution. *Electrochimica Acta,* 40 (12):1983–1995, 1995.

[96]   G. Strang and G.J. Fix. *An analysis of the finite element method.* Wellesley-Cambridge Press, 2008. ISBN 9780980232707.

[97]   R. Struijs, H. Deconinck, and P. L. Roe. *Fluctuation splitting schemes for the 2D Euler equations. In Lecture Series LS-1991-01.* von Karman Institute, SintGenesiusRode, Belgium, 1991.

[98]   K. Stüben. Algebraic multigrid (AMG): Experiences and comparisons. *Appl. Math. Comput.,* 13:419–452, 1983.

[99]   K. Stüben. Algebraic multigrid (AMG): An introduction with applications. *GMD Report,* 53, 1999.

[100]   K. Stüben. An introduction to algebraic multigrid. pages 413–532. In [106], 2001.

[101]   R. Tarjan. Depth-first search and linear graph algorithms. *SIAM Journal on Computing,* 1(2):146–160, 1972.

[102]   T.E. Tezduyar, S. Mittal, S.E. Ray, and R. Shih. Incompressible flow computations with stabilized bilinear and linear equal-order-interpolation velocity-pressure elements. *Comp. Meth. in Appl. Mechanics and Eng.,* pages 221–242, 1992.

[103]   P. Thum. Algebraic multigrid for Navier-Stokes equations - Studies on smoothing and coarsening. Master's thesis, Universität zu Köln, 2006.

[104]   P. Thum and T. Clees. Towards physics-oriented smoothing in algebraic multigrid for systems of partial differential equations arising in multi-ion transport and reaction models. *Numerical Linear Algebra with Applications,* 17(2-3):253–271. doi: 10.1002/nla.706,2010.

[105] P. Thum, T. Clees, G. Weyns, G. Nelissen, and J. Deconinck. Efficient algebraic multigrid for migration-diffusion-convection-reaction systems arising in electrochemical simulations. *Journal of computational physics*, 229(19): 7260–7276, 2010. doi: j.jcp.2010.06.011.

[106] U. Trottenberg, C. Oosterlee, and A. Schüller. *Multigrid*. Academic Press, 2001.

[107] S. Turek. On ordering strategies in a multigrid algorithm, 1997. URL `citeseer.ist.psu.edu/article/turek97ordering.html`.

[108] B. van den Bossche. *Numerical modelling and validation of electrochemical copper deposition processes for the production of printed circuits*. PhD thesis, Vrije Universiteit Brussel, Belgium, 1998.

[109] B. van den Bossche, L. Bortels, J. Deconinck, S. Vandeputte, and A. Hubin. Quasi-onedimensional steady-state analysis of multi-ion electrochemical systems at a rotating disc electrode controlled by diffusion, migration, convection and homogeneous reactions. *J. Electroanal. Chem.*, 397:35–44, 1995.

[110] P. Vaněk, J. Mandel, and M. Brezina. Algebraic multigrid on unstructured meshes. Technical Report UCD/CCM Report 34, Center for Computational Mathematics, University of Colorado at Denver, Denver(CO),USA, 1994.

[111] P. Vaněk, M. Brezina, and J. Mandel. Convergence of algebraic multigrid based on smoothed aggregation. *Numerische Mathematik*, 88:559–579, 2001.

[112] R. S. Varga. *Matrix Iterative Analysis*. Springer Berlin, 2009.

[113] W. Wan, T. Chan, and B. Smith. An energy-minimization interpolation for robust multigrid methods. *SIAM J. Sci. Comp.*, 21:1632–1649, 2000.

[114] N. P. Waterson. *Simulation of Turbulent Flow, Heat and Mass Transfer Using a Residual Distribution Approach*. PhD thesis, TU Delft, 2003.

[115] R. Webster. Algebraic multigrid and incompressible fluid flow. *Int J. Num. Meth. Fluids*, 53:669–690, 2007. Published online in Wiley InterScience (www.interscience.wiley.com).

[116] J. M. Weiss, J. P. Maruszewski, and W. A. Smith. Implicit solution of preconditioned Navier Stokes equations using algebraic multigrid. *AIAA J.*, 37:29–36, 1999.

List of Figures

List of Tables

214